Practical physics

PRACTICAL PHYSICS

G. L. SQUIRES
Lecturer in Physics at the University of Cambridge
and Fellow of Trinity College, Cambridge

THIRD EDITION

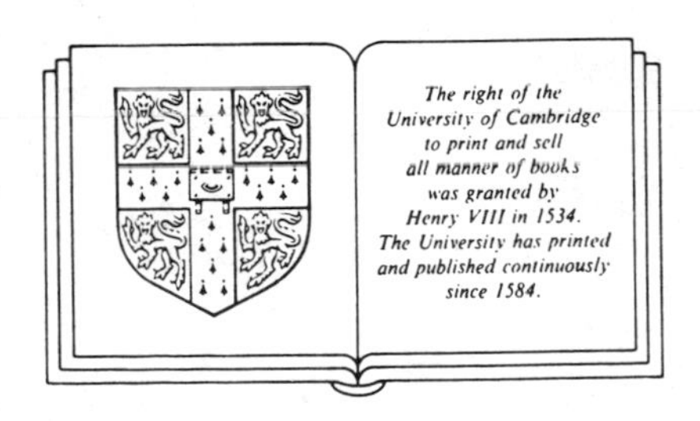

CAMBRIDGE UNIVERSITY PRESS
Cambridge
New York Port Chester
Melbourne Sydney

Published by the Press Syndicate of the University of Cambridge
The Pitt Building, Trumpington Street, Cambridge CB2 1RP
40 West 20th Street, New York, NY 10011, USA
10 Stamford Road, Oakleigh, Melbourne 3166, Australia

First published by McGraw-Hill Book Company (UK) Limited 1968
Third edition published by Cambridge University Press 1985
Reprinted 1988, (with revisions) 1989

Printed in Great Britain by J. W. Arrowsmith Ltd, Bristol

Library of Congress catalogue card number: 84-29196

British Library Cataloguing in Publication Data
Squires, G. L.
Practical physics. – 3rd ed.
1. Physics – Experiments
I. Title
530′.0724 QC33

ISBN 0 521 24952 X hard covers
ISBN 0 521 27095 2 paperback

CONTENTS

PART 2 EXPERIMENTAL METHODS

PART 3 RECORD AND CALCULATIONS

PREFACE TO THE THIRD EDITION

Despite major changes that have occurred in experimental methods since this book was first published, I believe that its object, namely to demonstrate the purposive and critical approach that should be made to all experimental work in physics, is as valid today as it was then. The basic outlook of the book has therefore been retained. However, a number of changes have been made in response to changes in calculating and experimental methods.

I have rewritten chapter 3, which gives the basic statistical theory for the estimation of the error in a single variable, and chapter 12, which deals with arithmetic, to take account of the widespread use of electronic calculators.

Secondly, I have made substantial changes in chapter 7, in which four experiments are analysed to show the art and craft of the experimenter. The first is based on the dc potentiometer, which at one time was a standard tool in research work. This is no longer so, but I have nevertheless retained the account, partly because the instrument has so many instructive features, and partly because it is readily available in schools and universities. However, to show how this type of electrical measurement is made nowadays in research work, I have followed with a description of a modern electronic method. In the previous editions, the chapter ended with an account of a precision measurement of the magnetic moment of the electron. This experiment has become so sophisticated as to make its description unsuitable for illustrative purposes. I have therefore replaced it by an account of a precision measurement of the gravitational acceleration *g*. Apart from the fact that the basic physics of the experiment is easier to understand, it contains several examples of elegance and ingenuity. Indeed, I have chosen the four experiments in the chapter because I feel they all have something to offer under these headings. Elegance and ingenuity cannot be taught, but I think examples are worth studying for their own sake, and, who knows, perhaps something may rub off on us.

Finally, I have taken the opportunity of bringing up to date some of the examples, exercises, references, and definitions of units.

The advent of microcomputers and microprocessors has produced major changes in experimental work. Detailed accounts of these devices are outside the scope of this book, but I would make one comment. There is no doubt that the use of computers is now an essential part of a student's training. But it would be unfortunate if this resulted in a significant reduction in the time spent doing actual experimental work. There is no substitute for the latter - not even reading this book.

I should like to thank Professor A. H. Cook, Dr P. J. Duffett-Smith, Dr R. H. Friend, Dr G. G. Lonzarich, and Dr A. T. Winter for helpful discussions, and for comments on the revised parts of the book.

G. L. SQUIRES
August 1984

PREFACE TO THE FIRST EDITION

Experimental physics has occupied some of the finest intellects in the history of man, but the fascination of the subject is not always apparent in an undergraduate course of practical work. This book is about experimental physics and it is intended for undergraduates, but it does not describe a systematic course of experiments, nor is it a handbook of experimental techniques. Instead, it sets out to demonstrate a certain outlook or approach to experimental work. It is intended as a companion to a general course of practical work. My aim is to make the student more critical of what he does and more aware of what can be done, and in this way to make the course more interesting and meaningful.

The book is in three parts. The first is on the statistical treatment of data. I have tried to give the statistical theory not as an exercise in mathematics but rather as a tool for experimental work. This is perhaps the most difficult part of the book, and the student should not worry if he does not grasp all the mathematical details at first. He should read through the chapters to get a general understanding – and then go ahead and use the results. He can always return and master the proofs at a later stage. The second part is on experimental methods. I discuss a selection of instruments, methods, and experiments with a view to showing the craft of the experimenter. The selection is arbitrary – though I have tried to illustrate the points with methods that are useful in themselves. The third part concerns such matters as keeping efficient records, getting arithmetic right, and writing good scientific English.

The examples have been kept fairly simple. Apart from the account of the measurement of the magnetic moment of the electron, the level of the material is roughly that of a first-year undergraduate course. But I think that a wider range of students – from intelligent sixth-formers to research students – could benefit from the experimental 'awareness' that the book is trying to foster.

The experiment to measure the magnetic moment of the electron is an advanced one and contains several ideas beyond those of an average first-year course. I have tried to give sufficient explanation to make it intelligible to someone in his second or third year who has had an introduction to quantum mechanics. The experiment is a rewarding one to study, but the whole account may be omitted at first reading without detriment to the understanding of the rest of the book.

I would like to thank Professor O. R. Frisch, Professor R. G. Chambers, Mr E. S. Shire, Dr J. Ashmead, Dr J. R. Waldram, Dr B. D. Josephson, and Dr N. J. B. A. Branson for reading the first draft of the book and making valuable suggestions for its improvement. I would also like to thank Messrs R. A. Bromley, R. G. Courtney, B. C. Hamshere, H. M. C. Rowe, B. Scruton, D. R. Weaver, and M. A. G. Willson who read parts of the first draft and made many useful comments from the user's point of view. Finally, I wish to express my indebtedness to all the undergraduates who have passed through the first-year Mechanics, Heat, and Optics Class at the Cavendish Laboratory in the past ten years. They have taught me much about errors - in every sense of the word.

G. L. SQUIRES
September 1967

1

The object of practical physics

This book is intended to help you to do practical physics at college or university: its aim is to make the laboratory work more useful and profitable. We may start by asking what is the object of laboratory work in a university physics course. There are several possible objects. Laboratory work may serve

(a) to demonstrate *theoretical ideas* in physics,
(b) to provide a familiarity with *apparatus*,
(c) to provide a training in *how to do experiments*.

Let us consider each of these in turn.

Seeing something demonstrated in practice is often a great help in understanding it. For example, interference in light is not an intuitive concept. The idea that two beams of light can cancel each other and give darkness takes a little swallowing, and most people find it helpful to be given a visual demonstration. A demonstration is useful for another reason – it gives an idea of orders of magnitude. The interference fringes are in general close together, which indicates that the wavelength of light is small compared with everyday objects. But the demonstration is no substitute for a proper explanation, which goes into the details of geometry and phase relations. So the first object, the demonstration of theoretical ideas, has a definite but limited usefulness.

The second object is perhaps more important, but it is necessary to say exactly what is meant by 'apparatus' in this context. In any practical course you will handle a number of instruments, such as oscilloscopes, timers, thermometers, and so on, and the experience you gain from using them should prove useful. However, if you eventually do scientific work of some kind, the range of instruments you could conceivably work with is enormous. No practical course could possibly teach you to use them all. What the course should do is to train you to use instruments *in general.* There is a certain attitude of mind that an experimenter should adopt when handling any instrument, and this the course should try to instil. But this is part of the third object which is the most important of all.

The phrase 'how to do experiments' may sound vague, so let us try to be more specific. The primary object - or set of objects - of practical physics is to train you to

(a) plan an experiment whose precision is appropriate to its purpose,
(b) be aware of and take steps to eliminate systematic errors in methods and instruments,
(c) analyse the results in order to draw correct conclusions,
(d) estimate the precision of the final result,
(e) record the measurements and calculations accurately, clearly, and concisely.

All this adds up to saying that the main object of a course in practical physics is to train you to be a competent experimenter. But the course can do still more. It can show the way physics works.

Physics is one of the natural sciences, that is to say, it is part of our attempt to understand the natural world. When we are confronted by a situation in the natural world, the way we proceed in physics is to select what we think are the essential features. For example, the Greeks saw that a moving body came to rest and they therefore said that a force is necessary to keep a body moving. Galileo and Newton observed the same phenomenon, but they said that the coming to rest of the body is not an essential feature of the situation. It depends on friction: in the absence of friction a body keeps moving. If we try to do an experiment to test this view, we find that we cannot completely eliminate friction or other retarding forces, but we can make such forces small, and the smaller we make them the farther the body goes before coming to rest. So it is reasonable to believe that in the limiting case of zero friction the motion will remain unchanged as stated in Newton's first law.

This is the way physics works. We select what we think are the essential features in an actual physical situation. From them we make a generalization, or theory, and from the theory, deductions. We test a deduction by doing an experiment. But the deduction refers to an idealized or theoretically simple situation. In order to test it we have to create this simple situation in the messy, complicated, natural world, which is often a difficult thing to do.

In lectures you are taught the theory of the subject. The physical world is described in terms of the features which the current theories say are essential. These features tend to be the only ones you hear about, and you may well come to feel that they constitute the entire world, instead of a specially selected part of it. Moreover, everything fits together so smoothly that you can easily lose sight of the genius and effort that went into creating the subject. The most effective antidote to this is going into the laboratory and seeing the complications of real life.

By doing practical physics, then, you learn at first hand some of the obstacles to testing a theory, to measuring what you want to measure and not something else, and you learn how to overcome them. But above all you get an insight into physics as a whole, into the way experiment and theory interact, which is the very essence of the subject.

PART 1

Statistical treatment of data

2

Introduction to errors

2.1 The importance of estimating errors

When we measure a physical quantity, we do not expect the value obtained to be exactly equal to the true value. It is important to give some indication of how close the result is likely to be to the true value, that is to say, some indication of the precision or reliability of the measurements. We do this by including with the result an estimate of its error. For example, we might measure the focal length f of a lens and give the final result as

$$f = (256 \pm 2)\ \text{mm}. \tag{2.1}$$

By this we mean that we expect the focal length to be somewhere in the range 254 to 258 mm. Equation (2.1) is really a probability statement. It means, not that we are *certain* that the value lies between the limits quoted, but that our measurements indicate that there is a certain *probability* of its doing so. In chapter 3 we shall make this statement more precise.

Estimates of errors are important because without them we cannot draw significant conclusions from the experimental results. Suppose, for example, we wish to find out whether temperature has an effect on the resistance of a coil of wire. The measured values of the resistance are

$$200{\cdot}025\ \Omega \text{ at } 10\ ^\circ\text{C}$$
$$200{\cdot}034\ \Omega \text{ at } 20\ ^\circ\text{C}.$$

Is the difference between these two values significant? Without knowing the errors we cannot say. If, for example, the error in each value of the resistance is 0·001 Ω, the difference is significant, whereas if the error is 0·010 Ω, then it is not.

Once the result of an experiment has been obtained it goes out into the world and becomes public property. Different people may make use of it in different ways. Some may use it in calculations for a practical end; others may want to compare it with a theoretical prediction. For example, an electrical engineer may want to know the resistivity of copper in order to design a transformer, while a solid state physicist may want to know the same quantity to test a theory of electrons in metals. Whatever use a person makes of an experimental result, he will want to know whether it is sufficiently precise for his purpose. If he has drawn some conclusions from the result, he will want to know how much confidence to place in them. To answer such questions, an estimate of the error in the result is necessary, and it is the responsibility of the experimenter to provide it.

Now although the experimenter may not be able to foresee all the possible uses of his results, he should be aware of some of them. No experiment should be done in a vacuum – at least not in an intellectual one. If the experiment is being done to test a theory, the experimenter should have some idea how precise the result needs to be in order to provide a useful comparison with the theoretical prediction. So the idea of using error estimates to draw conclusions from the results of an experiment applies also in reverse. That is to say, the purpose of the experiment often determines the error that can be tolerated, and this in turn may have an important influence on the experimental procedure.

It might be thought that every experiment should be done as precisely as possible, but that is an unrealistic point of view. Life is finite, so are the experimenter's resources, and so also, unless he is a genius, is his capacity for taking pains. Therefore it is important to plan and carry out the experiment so that the precision of the final answer is appropriate to the ultimate object of the experiment. Suppose, in the example given at the beginning of this section, that we are only interested in the resistance of the coil because we want to use it as a standard resistance in the temperature range 10 °C to 20 °C, and that the precision required is 1 part in 10 000. A measurement of the resistance with an error of 0·010 Ω would then be quite adequate, and to strive to reduce the error to 0·001 Ω would be a waste of time. On the other hand, to measure the resistance to only 0·05 Ω would be even worse because the measurements would be useless for their purpose.

Just as the final result of an experiment should be obtained to an appropriate degree of precision, so also should the values of the various measured quantities within the experiment. Few experiments are so simple that the final quantity is measured directly. We usually have to measure several primary quantities and bring the results together in order to obtain the quantity required. The errors in the primary quantities determine that in the final result. In general the primary errors contribute different amounts to the final error, and the latter is minimized if the finite resources of time, apparatus, and patience available are concentrated on reducing those errors that contribute most to the final error.

So we see that the idea of errors is not something of secondary or peripheral interest in an experiment. On the contrary, it is related to the purpose of the experiment, the method of doing it and the significance of the results.

2.2 Systematic and random errors

Errors may be divided into two kinds, systematic and random. A *systematic* error is one which is constant throughout a set of readings.* A *random* error is one which varies and which is equally likely to be positive or negative.

Random errors are always present in an experiment and, in the absence of systematic errors, they cause successive readings to spread about the true value of the quantity – Fig. 2.1a. If in addition a systematic error is present, the readings spread, not about the true value, but about some displaced value – Fig. 2.1b.

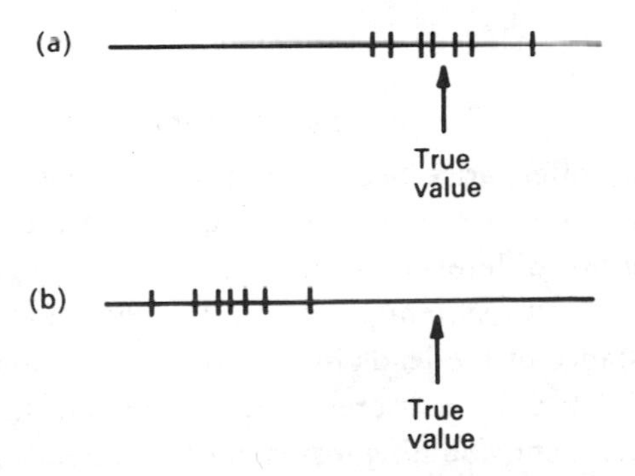

Fig. 2.1. Set of measurements (a) with random errors only and (b) with systematic plus random errors. Each point indicates the result of a measurement.

Suppose that the period of a pendulum is measured by means of a stopclock, and the measurement is repeated many times. Errors in starting and stopping the clock, in estimating the scale divisions, small irregularities in the motion of the pendulum, all these cause variations in the results of successive measurements and may be regarded as random

* This definition is actually too restrictive – some systematic errors are not constant. But in order to give the basic ideas we restrict the discussion here to the simple case. More general cases are considered in chapter 8.

errors. If no other errors are present, some results will be too high and others too low. But if, in addition, the clock is running slow, all the results will be too low. This is a systematic error.

It should be noticed that systematic and random errors are *defined* according to whether they produce a systematic or random effect. So we cannot say that a certain source of error is inherently systematic or random. Returning to the example, suppose that every time we measure the period we use a different clock. Some clocks may be running fast and others slow. But such inaccuracies now produce a random error.

Again, some sources of error may give rise to both systematic and random effects. For example, in operating the clock we might not only start and stop it in a slightly irregular manner in relation to the motion of the pendulum, thus producing a random error, but we might also have a tendency to start it too late and stop it too early, which would give rise to a systematic error.

It is convenient to make a distinction between the words *accurate* and *precise* in the context of errors. Thus a result is said to be *accurate* if it is relatively free from systematic error, and *precise* if the random error is small.

2.3 Systematic errors

Systematic errors often arise because the experimental arrangement is different from that assumed in the theory, and the correction factor which takes account of this difference is ignored. It is easy to give examples of effects that may lead to systematic error. Thermal emfs in a resistance bridge, the resistance of the leads in a platinum thermometer, the effect of the exposed stem in a mercury thermometer, heat losses in a calorimetric experiment, counting losses due to the dead-time in a particle counter are but a few. Another common source of systematic error is the one mentioned earlier - inaccurate apparatus.

Random errors may be detected by repeating the measurements. Furthermore, by taking more and more readings we obtain from the arithmetic mean a value which approaches more and more closely to the true value. Neither of these points is true for a systematic error. Repeated measurements with the same apparatus neither reveal nor do they eliminate a systematic error. For this reason systematic errors are potentially more dangerous than random errors. If large random errors are present in an experiment, they will manifest themselves in a large value of the final quoted error. Thus everyone is aware of the imprecision of the result, and no harm is done - except possibly to the ego of the experimenter when no one takes notice of his results. However, the concealed presence of a systematic error may lead to an apparently reliable result, given with a small estimated error, which is in fact seriously wrong.

A classic example of this was provided by Millikan's oil-drop experiment to measure e, the elementary charge. In this experiment it is necessary to know the viscosity of air. The value used by Millikan was too low, and as a result the value he obtained for e was

$$e = (1{\cdot}591 \pm 0{\cdot}002) \times 10^{-19}\ \mathrm{C}.$$

This may be compared with the present value (Cohen and Taylor 1987)

$$e = (1{\cdot}602\,177\,3 \pm 0{\cdot}000\,000\,5) \times 10^{-19}\ \mathrm{C}.$$

Up till 1930, the values of several other atomic constants, such as the Planck constant and the Avogadro constant, were based on Millikan's value for e and were consequently in error by more than $\frac{1}{2}\%$.

Random errors may be estimated by statistical methods, which are discussed in the next two chapters. Systematic errors do not lend themselves to any such clear-cut treatment. Your safest course is to regard them as effects to be discovered and eliminated. There is no general rule for doing this. It is a case of thinking about the particular method of doing an experiment and of always being suspicious of the apparatus. We shall try to point out common sources of systematic error in this book, but in this matter there is no substitute for experience.

3

Treatment of a single variable

3.1 Introduction

Suppose we make a set of measurements, free from systematic error, of the same quantity. The individual values x_1, x_2, etc., vary owing to random errors, and the mean value $\bar{x}$ (i.e. the arithmetic average) is taken as the best value of the quantity. But, unless we are lucky, $\bar{x}$ will not be equal to the true value of the quantity, which we denote by X. The question we are going to consider is how close we expect $\bar{x}$ to be to X. Of course we do not know the *actual* error in $\bar{x}$. If we did, we would correct $\bar{x}$ by the required amount and get the right value X. The most we can do is to say that there is a certain probability that X lies within a certain range centred on $\bar{x}$. The problem then is to calculate this range for some specified probability.

A clue to how we should proceed is provided by the results shown in Fig. 3.1. On the whole, for the results in Fig. 3.1a we would expect X to be fairly close to $\bar{x}$; whereas for those in Fig. 3.1b we would not be greatly surprised if there were quite a large difference. In other words, the larger the spread in the results, the larger we would expect the error

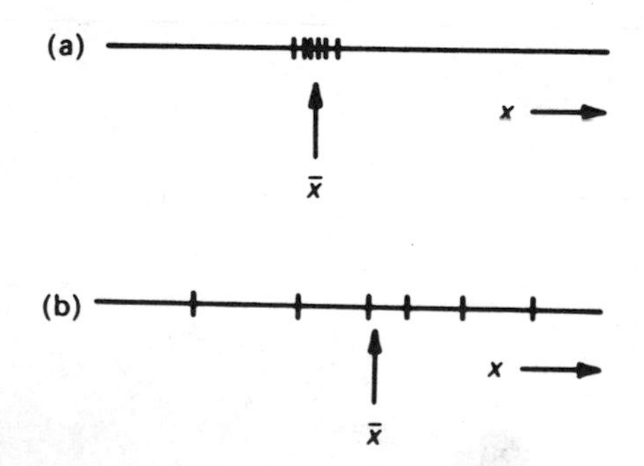

Fig. 3.1. Results of successive measurements of the same quantity. The mean $\bar{x}$ is expected to be closer to the true value for set (a) than for set (b).

Table 3.1. *Measurements of the resistance R of a coil*

R/Ω	R/Ω
4·615	4·613
4·638	4·623
4·597	4·659
4·634	4·623

in $\bar{x}$ to be. The whole of the present chapter is concerned with putting this qualitative statement on a firm quantitative basis. We assume throughout that no systematic errors are present.

3.2 Set of measurements

Denote the values of n successive measurements of the same quantity by

$$x_1, x_2, \ldots, x_n. \tag{3.1}$$

The number n is not necessarily large and in a typical experiment might be in the range 5 to 10. The mean is

$$\bar{x} = \frac{1}{n} \sum x_i. \tag{3.2}$$

(Whenever the symbol $\sum$ appears in the present chapter, the summation is to be taken from $i = 1$ to $i = n$.)

To fix our ideas let us consider a specific experiment in which the resistance of a coil is measured on a bridge. The measurement is made $n = 8$ times. The results are listed in Table 3.1. The mean of these values is 4·625 Ω. We require a quantity that gives a measure of the spread in the 8 values, from which we shall estimate the error in the mean. To define such a quantity we need to introduce the idea of a *distribution* – one of the basic concepts in the theory of statistics.

3.3 Distribution of measurements

(a) Introduction. Although we have only n actual measurements, we *imagine* that we go on making the measurements so that we end up with a very large number N. We may suppose N is say 10 000 000. (Since we are not actually making the measurements, expense is no object.) We call this hypothetical set of a very large number of readings a *distribution*.

The basic idea to which we shall constantly return is that *our actual set of n measurements is a random sample taken from the distribution of N measurements.*

We may represent any set of measurements by a *histogram.* This is true whether it be a set of a small number n of measurements or a distribution of a large number N. To construct a histogram we simply divide the range of measured values into a set of equal intervals and count how many times a measurement occurs in each interval. The width of the intervals is arbitrary and is chosen for convenience. Figure 3.2 shows a histogram for the measurements in Table 3.1.

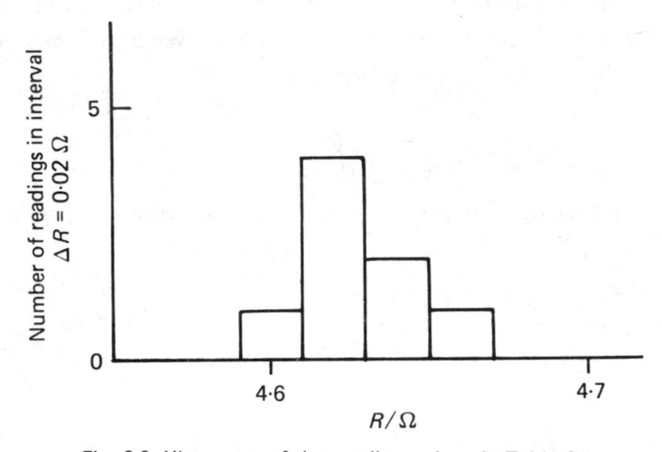

Fig. 3.2. Histogram of the readings given in Table 3.1.

This histogram has a jagged appearance because it represents only a few values. However, suppose we represent a distribution in this way. The number of measurements N is so large that we may make the width of thc intervals very small and – provided the measuring instrument gives sufficiently fine readings – still have an appreciable number of readings in each interval. If we plot, instead of the histogram itself, the *fraction* of the N readings in each interval as a function of the value of the measurement, we shall get a smooth curve. We may then define a function $f(x)$, known as the *distribution function,* whose significance is that $f(x)\,\mathrm{d}x$ is the fraction of the N readings that lie in the interval x to $x+\mathrm{d}x$. In other words, $f(x)\mathrm{d}x$ is the *probability* that a single measurement taken at random from the distribution will lie in the interval x to $x+\mathrm{d}x$. We shall not specify the exact form of $f(x)$ at this stage but we expect a typical distribution function to look something like that shown in Fig. 3.3.

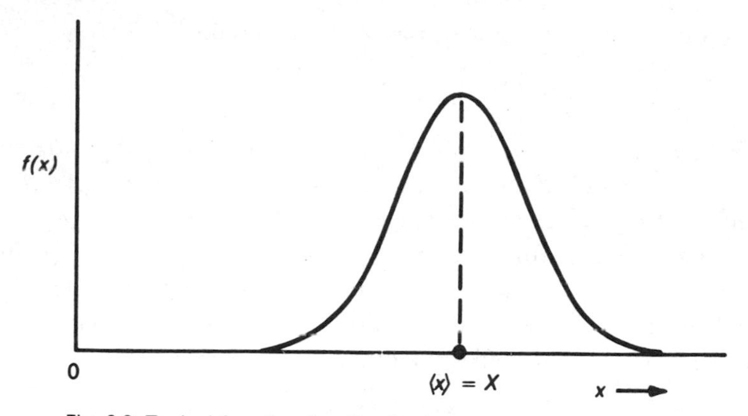

Fig. 3.3. Typical function for distribution of measurements.

From its definition $f(x)$ satisfies the relation

$$\int_{-\infty}^{\infty} f(x)\,\mathrm{d}x = 1. \tag{3.3}$$

Notice the infinite limits in the integral. We do not expect any measurements with values greatly different from the true value X in an actual experiment. In particular, many quantities are of a nature such that negative values are impossible. Therefore any function $f(x)$ that we use to represent a distribution must become very small as the difference between x and X becomes large. For such functions no difficulty arises from the infinite limits, and they are taken for mathematical convenience.

We shall use the symbol $\langle\,\rangle$ to denote an average over all the measurements in the distribution. An important average is the *mean* of the distribution

$$\langle x\rangle = \int_{-\infty}^{\infty} xf(x)\,\mathrm{d}x. \tag{3.4}$$

Since the number of measurements in the distribution is very large, and they are free from systematic error, $\langle x\rangle$ may be taken as equal to the true value X.

(b) Standard error in a single observation. The error in a measurement with value x is

$$e = x - X. \tag{3.5}$$

The rms (root-mean-square) value of e for all the measurements in the distribution is denoted by σ and is known as the *standard deviation of* *the distribution.** Thus σ is defined by the equation

$$\sigma^2 = \langle e^2 \rangle = \int_{-\infty}^{\infty} (x-X)^2 f(x)\,dx. \tag{3.6}$$

The standard deviation is a measure of the spread of the distribution, i.e. of the scatter of the measurements. A distribution representing a precise set of measurements will be highly peaked near $x = X$ and will have a small value for σ; while one representing an imprecise set will have a large scatter about X and a large value for σ (Fig. 3.4). We take the quantity σ as our measure of the error in a single observation, and it is therefore also referred to as the *standard error in a single observation.*

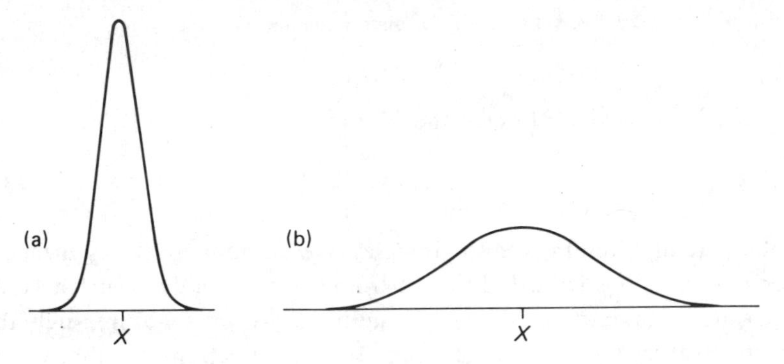

Fig. 3.4. Distribution function $f(x)$ for (a) a precise set of measurements (small value of σ), and (b) an imprecise set of measurements (large value of σ). Note that the area under the two curves is the same because both functions satisfy the relation (3.3).

(c) Standard error in the mean. We now proceed to define a quantity that specifies the error in the mean of a set of n measurements.

Let us go back to the set of 8 measurements given in Table 3.1. We have said that they are to be regarded as a random sample taken from the distribution of single measurements. Imagine the distribution to be represented by a bin containing N balls, each of which is labelled with the value of a single measurement. The set of measurements in Table 3.1 may be regarded as the result of drawing 8 balls at random from the bin. Suppose now we have a second bin, initially empty, and a supply of blank balls. We look at our 8 measurements, calculate their mean, write the answer on a blank ball and put it in the second bin. We put the set of 8 single-measurement balls back into the first bin, stir the contents,

* The quantity σ^2 is known as the *variance* of the distribution.

and take out, at random, a second set of 8 balls. We again calculate the mean of the 8 values, write it on a blank ball, and put it in the second bin. We continue the process a large number of times, always drawing the same number, 8, of balls from the first bin. We end up with a large number of values in the second bin, which represents another distribution, namely a distribution of means of 8 measurements.

We denote the standard deviation of this distribution by σ_m. We take it as our measure of the error in the mean of a single set of n measurements, and it is therefore known as the *standard error in the mean.*

To sum up, σ is the standard deviation of the distribution of single measurements, and σ_m is the standard deviation of the distribution of the means of sets of measurements, each set containing the same number n of single measurements. σ represents the error in a single measurement, and σ_m represents the error in the mean of n measurements.

(d) Relation between σ and σ_m. There is a simple relation between σ and σ_m which we now prove. Consider one set of n measurements $x_1, \ldots, x_n$. The error in the ith reading is

$$e_i = x_i - X, \tag{3.7}$$

where X is the true value of the quantity, which is of course unknown. The error in the mean is

$$E = \bar{x} - X = \left(\frac{1}{n}\sum x_i\right) - X = \frac{1}{n}\sum (x_i - X) = \frac{1}{n}\sum e_i. \tag{3.8}$$

Therefore

$$E^2 = \frac{1}{n^2}\sum e_i^2 + \frac{1}{n^2}\sum_{i}\sum_{\substack{j\\ i\neq j}} e_i e_j. \tag{3.9}$$

This is for a single set of n measurements. Now imagine that, as before, we take a large number of sets, each set consisting of the same number n of single measurements. Each set will have its own set of values for $e_1, \ldots, e_n$, and its own value of E. Equation (3.9) will be true for each set. We add the equations for all the sets and then divide by the number of sets, that is to say, we average (3.9) over all the sets. The average of $\sum e_i^2$ is $n\langle e^2\rangle$. The average of each term in the double sum is zero, because the errors e_i and e_j are independent, and the average of each is zero. We therefore arrive at the result

$$\langle E^2\rangle = \frac{1}{n}\langle e^2\rangle. \tag{3.10}$$

By definition

$$\sigma_{\mathrm{m}}^2 = \langle E^2 \rangle \quad \text{and} \quad \sigma^2 = \langle e^2 \rangle. \tag{3.11}$$

Equation (3.10) thus becomes

$$\sigma_{\mathrm{m}} = \frac{\sigma}{\sqrt{n}}, \tag{3.12}$$

i.e. *the standard error in the mean of n observations is* $1/\sqrt{n}$ *times the standard error in a single observation.*

The value of σ depends only on the precision of the individual measurements and is independent of their number; whereas the value of σ_{m} can be reduced by increasing n. However, since σ_{m} decreases only as $1/\sqrt{n}$, it becomes more and more unprofitable to take readings of the same quantity. Rather we should try to reduce σ_{m} by reducing σ, i.e. by taking a more precise set of readings in the first place.*

3.4 Estimation of σ and σ_{m}

(a) Standard method. We have defined the quantities σ and σ_{m} that we are going to take as our measures of the error in a single measurement and in the mean. It remains to show how we can calculate, or more correctly how we can best estimate, them from the actual measurements. We need to estimate only one of them, because we can then use (3.12) to obtain the other.

The best estimate of σ is $[(1/n)\sum e_i^2]^{\frac{1}{2}}$, but the errors e_i come from the true value X and hence are not known. A way round the difficulty is provided by working in terms of *residuals.*

The residual d_i of the ith measurement is defined by

$$d_i = x_i - \bar{x}. \tag{3.13}$$

Unlike the error, the residual is a known quantity. We denote the rms value of the n residuals by s, i.e.

$$s^2 = \frac{1}{n} \sum d_i^2. \tag{3.14}$$

The quantity s is called the *standard deviation of the sample.* From (3.7) and (3.8)

$$x_i - \bar{x} = e_i - E. \tag{3.15}$$

* A splendid example of this approach is to be found in the experiment to measure g precisely – see section 7.4 (d).

Therefore

$$s^2 = \frac{1}{n}\sum (x_i - \bar{x})^2 = \frac{1}{n}\sum (e_i - E)^2$$

$$= \frac{1}{n}\sum e_i^2 - 2E\frac{1}{n}\sum e_i + E^2$$

$$= \frac{1}{n}\sum e_i^2 - E^2. \tag{3.16}$$

This is for one set of n measurements. As before, we take the average of this equation over a large number of sets in the distribution and obtain the result

$$\langle s^2 \rangle = \sigma^2 - \sigma_m^2. \tag{3.17}$$

From (3.12) and (3.17) we have

$$\sigma^2 = \frac{n}{n-1}\langle s^2 \rangle, \tag{3.18}$$

and

$$\sigma_m^2 = \frac{1}{n-1}\langle s^2 \rangle. \tag{3.19}$$

The quantity $\langle s^2 \rangle$ is not known. Our best estimate of it is s^2, obtained by evaluating (3.14). Substituting this value and taking the square root gives

$$\sigma \approx \left(\frac{n}{n-1}\right)^{\frac{1}{2}} s, \tag{3.20}$$

$$\sigma_m \approx \left(\frac{1}{n-1}\right)^{\frac{1}{2}} s. \tag{3.21}$$

We thus have estimates of σ and σ_m in terms of quantities that are known.*

(*b*) *Worked example.* As an example we show how σ and σ_m are estimated for the set of measurements in Table 3.1, which are listed again in the first column of Table 3.2. The first step is to calculate the mean, which is 4·625 Ω. From the mean we calculate the residual of each measurement. For example, the residual of the first measurement is

$$d_1 = (4{\cdot}615 - 4{\cdot}625)\ \Omega = -10\ \text{m}\Omega. \tag{3.22}$$

The residuals and their squares are listed in the second and third columns

* The symbol $\approx$ signifies that (3.20) and (3.21) are not strictly equations. The values of the right-hand sides depend on the particular set of n measurements and are not in general exactly equal to σ and σ_m – see section 3.7.

Table 3.2. *Estimation of σ and σ_m for the measurements in Table 3.1*

	Resistance R/Ω	Residual $d/\text{m}\Omega$	$(d/\text{m}\Omega)^2$
	4·615	−10	100
	4·638	13	169
	4·597	−28	784
	4·634	9	81
	4·613	−12	144
	4·623	−2	4
	4·659	34	1156
	4·623	−2	4
mean =	4·625		sum = 2442

of Table 3.2. Then

$$s^2 = \frac{1}{n}\sum d_i^2 = \frac{2442}{8} \times 10^{-6}\,\Omega^2, \qquad s = 0{\cdot}017\,\Omega, \tag{3.23}$$

$$\sigma \approx \left[\frac{n}{n-1}\right]^{\frac{1}{2}} s = \left(\frac{8}{7}\right)^{\frac{1}{2}} \times 0{\cdot}017 = 0{\cdot}019\,\Omega, \tag{3.24}$$

$$\sigma_m = \frac{\sigma}{\sqrt{n}} \approx \frac{0{\cdot}019}{\sqrt{8}} = 0{\cdot}007\,\Omega. \tag{3.25}$$

The result of a set of measurements is quoted as $\bar{x} \pm \sigma_m$. So in the present case our best estimate of the resistance of the coil is

$$R = 4{\cdot}625 \pm 0{\cdot}007\,\Omega. \tag{3.26}$$

(c) *Programmed calculator.* Calculators, programmed to calculate σ, use the standard method, but they do not evaluate s from (3.14) because that requires the value of $\bar{x}$, which is not known until all the numbers are fed in. However, there is an alternative expression for s that avoids the difficulty. From (3.2), (3.13) and (3.14) we have

$$\begin{aligned} s^2 &= \frac{1}{n}\sum (x_i - \bar{x})^2 \\ &= \frac{1}{n}\left[\sum x_i^2 - 2\bar{x}\sum x_i + n\bar{x}^2\right] \\ &= \frac{1}{n}\sum x_i^2 - \left(\frac{1}{n}\sum x_i\right)^2. \end{aligned} \tag{3.27}$$

Combining this with (3.20) gives

$$\sigma \approx \left(\frac{1}{n-1}\right)^{\frac{1}{2}} \left[\sum x_i^2 - \frac{1}{n}(\sum x_i)^2\right]^{\frac{1}{2}}. \tag{3.28}$$

This is the expression used by a calculator programmed to evaluate standard deviations. As the numbers x_i are fed in, the calculator accumulates the values of $\sum x_i^2$ and $\sum x_i$. It then uses (3.2) and (3.28) to evaluate $\bar{x}$ and σ. Similarly

$$\sigma_{\mathrm{m}} \approx \left[\frac{1}{n(n-1)}\right]^{\frac{1}{2}} \left[\sum x_i^2 - \frac{1}{n}(\sum x_i)^2\right]^{\frac{1}{2}}. \tag{3.29}$$

(d) Deviations from a general value of x. Before leaving this section we prove one further result. Suppose that instead of taking the deviations of the readings from the mean $\bar{x}$, as we have done so far, we take the deviations from a different value of x. Denote by $S(x)$ the rms deviation of the readings taken from x, so that

$$[S(x)]^2 = \frac{1}{n}\sum (x_i - x)^2. \tag{3.30}$$

Combining this equation with (3.27) we have

$$\begin{aligned} [S(x)]^2 - s^2 &= \frac{1}{n}\sum [(x_i - x)^2 - (x_i - \bar{x})^2] \\ &= \frac{1}{n}\sum (x^2 - 2x_i x + 2x_i\bar{x} - \bar{x}^2) \\ &= x^2 - 2\bar{x}x + 2\bar{x}^2 - \bar{x}^2 = (x - \bar{x})^2, \end{aligned}$$

i.e.

$$[S(x)]^2 = s^2 + (x - \bar{x})^2. \tag{3.31}$$

This demonstrates an important result. For a given set of readings the sum of the squares of the deviations is a minimum when the deviations are taken from the mean of the set. That is the reason why s^2 is not an unbiased estimate of σ^2. It is slightly too small, as (3.18) shows.

3.5 The Gaussian distribution

We have not yet specified the exact form of the distribution function $f(x)$. The results derived so far are therefore independent of the distribution. However, to make further progress we need a specific function, and the one we shall take is

$$f(x) = \frac{1}{\sqrt{(2\pi)}}\frac{1}{\sigma}\exp[-(x - X)^2/2\sigma^2]. \tag{3.32}$$

This function, specified by the two constants X and σ, is known as a *Gaussian*, and the distribution as a *Gaussian* or *normal* distribution. Its shape is shown in Fig. 3.5.

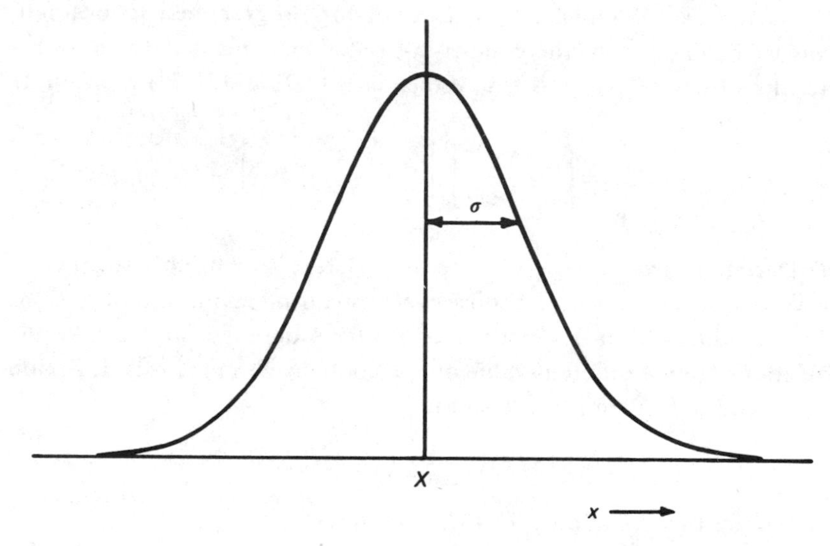

Fig. 3.5. The Gaussian distribution function. The points of inflexion are at $x = X \pm \sigma$.

Later in the chapter we mention the reasons for choosing the Gaussian, but at this stage we merely note that the function

(i) is symmetric about X,
(ii) has its maximum value at X,
(iii) tends rapidly to zero as $|x - X|$ becomes large compared with σ.

Clearly these are reasonable properties for a function representing a distribution of measurements containing only random errors.

We show below that the constant σ in (3.32) is in fact the standard deviation of the function – hence the choice of the symbol. The purpose of the multiplying factor

$$\frac{1}{\surd(2\pi)}\frac{1}{\sigma}$$

is to make $f(x)$ satisfy (3.3). We can see that it does so from the value of the first integral in Table 3.3. (The results in the table are proved in Appendix B.) Put $X = 0$, which does not affect the argument. Then

$$\int_{-\infty}^{\infty} f(x)\,\mathrm{d}x = \frac{1}{\surd(2\pi)}\frac{1}{\sigma}\int_{-\infty}^{\infty} \exp(-x^2/2\sigma^2)\,\mathrm{d}x = 1. \qquad (3.33)$$

Table 3.3. *Some useful integrals for the Gaussian distribution*

$$\int_{-\infty}^{\infty} \exp(-x^2/2\sigma^2)\,\mathrm{d}x = \surd(2\pi)\sigma$$

$$\int_{-\infty}^{\infty} x^2 \exp(-x^2/2\sigma^2)\,\mathrm{d}x = \surd(2\pi)\sigma^3$$

$$\int_{-\infty}^{\infty} x^4 \exp(-x^2/2\sigma^2)\,\mathrm{d}x = 3\surd(2\pi)\sigma^5$$

The standard deviation of the function in (3.32) is obtained from the second integral in Table 3.3. We continue with $X=0$. By definition

$$\begin{aligned}(\text{standard deviation})^2 &= \int_{-\infty}^{\infty} x^2 f(x)\,\mathrm{d}x \\ &= \frac{1}{\surd(2\pi)}\frac{1}{\sigma}\int_{-\infty}^{\infty} x^2 \exp(-x^2/2\sigma^2)\,\mathrm{d}x \\ &= \sigma^2. \end{aligned} \tag{3.34}$$

It is readily verified that the points of inflexion of the function $\exp(-x^2/2\sigma^2)$ occur at $x=\pm\sigma$, a convenient result for relating the standard deviation to the shape of a Gaussian.

3.6 The integral function

Suppose we have a symmetric distribution represented by a function $f(x)$ for which $X=0$. We can ask what fraction $\phi(x)$ of the measurements lie within the range $-x$ to x. Since $f(x)\,\mathrm{d}x$ is by definition the fraction of readings between x and $x+\mathrm{d}x$, $\phi(x)$ is given by

$$\phi(x) = \int_{-x}^{x} f(y)\,\mathrm{d}y. \tag{3.35}$$

We call $\phi(x)$ the *integral function* of the distribution. It is equal to the shaded area in Fig. 3.6 divided by the total area under the curve.

For a Gaussian distribution with standard deviation σ

$$\phi(x) = \frac{1}{\surd(2\pi)}\frac{1}{\sigma}\int_{-x}^{x} \exp(-y^2/2\sigma^2)\,\mathrm{d}y. \tag{3.36}$$

The function $\phi(x)$ depends on σ. It is convenient to have one table of

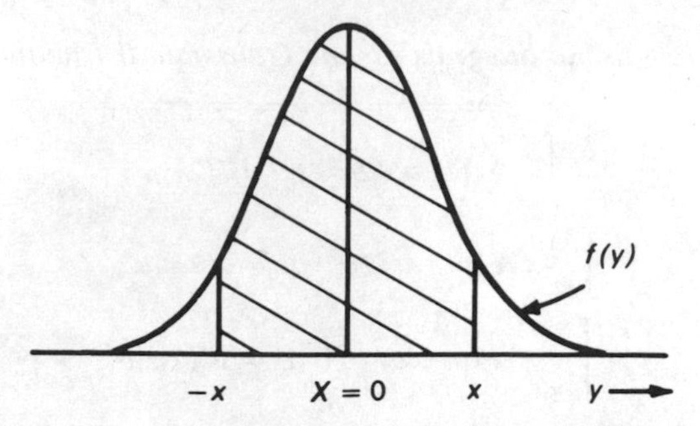

Fig. 3.6. $\phi(x)$, the fraction of measurements within $\pm x$, is the ratio of the shaded area to the total area under the distribution function $f(y)$.

values that can be used for all values of σ. The variable is therefore changed to $t = y/\sigma$. Put $z = x/\sigma$. Then

$$\phi(z) = \sqrt{\left(\frac{2}{\pi}\right)} \int_0^z \exp(-t^2/2)\,\mathrm{d}t. \tag{3.37}$$

The function $\phi(z)$ must be evaluated by numerical methods. It is tabulated in Appendix A and plotted in Fig. 3.7.

A few selected values of $\phi(z)$ are given in Table 3.4. The numbers in the third column are worth remembering. We see that about two out of

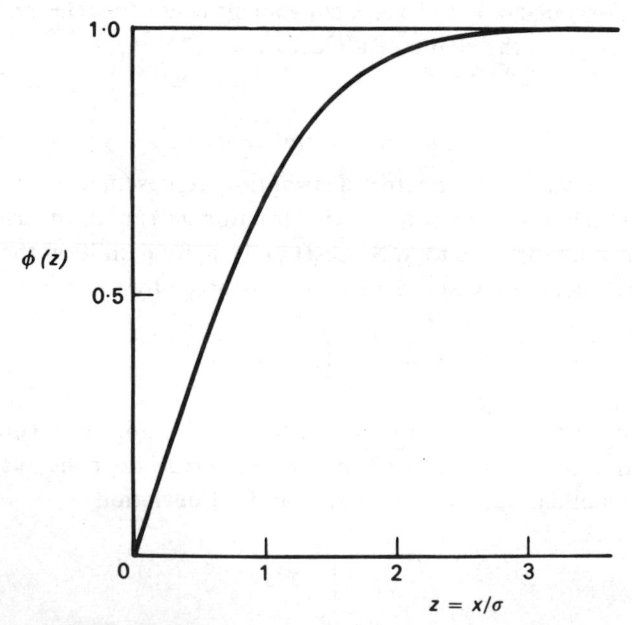

Fig. 3.7. The integral function $\phi(z)$ for the Gaussian distribution.

Table 3.4. *Values of the Gaussian integral function*

$z = x/\sigma$	$\phi(z)$	Approximate fraction of readings outside z value
0	0	1 out of 1
1	0·683	3
2	0·954	20
3	0·9973	400
4	0·99994	16 000

three observations lie within $\pm\sigma$. About 1 in 20 of the observations lie outside 2σ, and about 1 in 400 outside 3σ.

These results provide a quantitative basis for the statement that σ is a measure of the spread of the observations. They also provide a check that σ has been estimated correctly. For a set of readings with mean $\bar{x}$, roughly two-thirds of the readings should lie in the range $\bar{x} \pm \sigma$. We can also apply the results to the interpretation of σ_{m}, remembering that σ_{m} is the standard deviation of the distribution of means of which $\bar{x}$ is a member. Thus, when we quote the result of the measurements as $\bar{x} \pm \sigma_{\mathrm{m}}$, the implication is that, in the absence of systematic error, the probability that the true value of the quantity lies in the quoted range is roughly two-thirds.

In addition to σ, another quantity sometimes used to specify the error in the measurements is the so-called *probable error.* It is defined as the value of x such that one-half of the readings lie within x of the true value. For the Gaussian distribution the probable error is equal to $0{\cdot}67\sigma$. There is little point in having two similar measures of the error, and it is obviously convenient to settle on one of them. Though its significance is easy to appreciate, the probable error is not a very fundamental quantity, and the error commonly quoted nowadays is the standard error. We use it throughout the present book and only mention the probable error because it may be encountered in older books and papers.

3.7 The error in the error

It was stated in section 3.4 that the best estimate of $\langle s^2 \rangle$ is provided by s^2. However, s^2 is just the value we happen to get from the particular set of n readings. It is of interest to know how s^2 varies from one set to another. The error in s^2 is

$$u = s^2 - \langle s^2 \rangle. \qquad (3.38)$$

In Appendix C we show that for a Gaussian distribution the mean value of u^2 (taken from a large number of sets of n readings) is $[2/(n-1)]\langle s^2 \rangle^2$. Hence the fractional standard deviation of s^2 is $[2/(n-1)]^{\frac{1}{2}}$, and, provided n is fairly large, the fractional standard deviation of s is approximately half this value, i.e. $1/(2n-2)^{\frac{1}{2}}$.

The quantity $1/(2n-2)^{\frac{1}{2}}$ is plotted against n in Fig. 3.8. The result provides a salutary warning against elaborate calculations of errors. Notice for example that for $n = 9$, a not insignificant number of readings, the error estimate is only good to 1 part in 4.

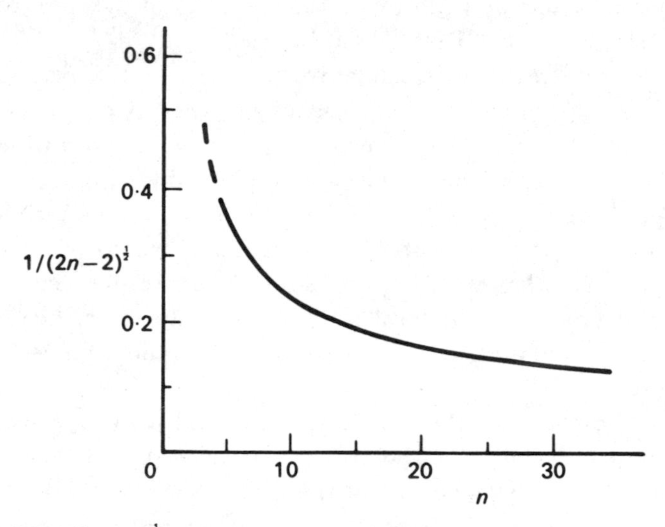

Fig. 3.8. $1/(2n-2)^{\frac{1}{2}}$, representing the fractional standard deviation of s, plotted against n, the number of measurements.

3.8 Range method of estimating σ and σ_m

The standard method of estimating σ and σ_m (section 3.4) is via s, the standard deviation of the sample. However, if you do not have a calculator programmed to calculate standard deviations, the expressions for s – (3.14) and (3.27) – are tedious to evaluate. Fortunately there are other methods of estimating σ and σ_m, which are easier to use, and are almost as reliable as the standard method. The simplest of these is the *range method.* It assumes that the distribution function is Gaussian.

Let r be the difference between the largest and smallest value in a set of n readings. Then an estimate of σ is given by the formula

$$\sigma \approx \frac{r}{v_n}, \tag{3.39}$$

where v_n is a number which may be calculated from statistical theory (Guest 1961, p. 44). It turns out that for n up to about 12, an adequate approximation for v_n is $\sqrt{n}$. So the method reduces to the simple formula

$$\sigma \approx \frac{r}{\sqrt{n}}. \tag{3.40}$$

Remember that σ is independent of n. Formula (3.40) therefore implies that the average value of the range r of a set of n readings is proportional to $\sqrt{n}$.

Applying the range method to the set of readings in Table 3.1 we have

$$r = 4{\cdot}659 - 4{\cdot}597 = 0{\cdot}062\ \Omega. \tag{3.41}$$

Therefore

$$\sigma \approx \frac{0{\cdot}062}{\sqrt{8}} = 0{\cdot}022\ \Omega, \tag{3.42}$$

which may be compared with $\sigma \approx 0{\cdot}019\ \Omega$ given by the standard method.

If we combine (3.40) with the relation

$$\sigma_m = \frac{\sigma}{\sqrt{n}}, \tag{3.12}$$

we have

$$\sigma_m \approx \frac{r}{n}, \tag{3.43}$$

which is a particularly simple formula for estimating σ_m. The formula does *not*, of course, imply that σ_m is proportional to $1/n$. Since r is proportional to $\sqrt{n}$, the formula implies correctly that σ_m is proportional to $1/\sqrt{n}$.

For values of n up to 12, the fractional standard deviations of the estimates of σ given by the range method are only about 10% higher than those given by the standard method. But for n greater than 12 the range estimates become increasingly unreliable. Also $\sqrt{n}$ becomes an increasingly poor approximation for v_n. ($\sqrt{n}$ becomes greater than v_n, so the range method tends to underestimate σ.) However, for values of n up to about 12, the range method of estimating σ and σ_m is quite adequate for most purposes.

3.9 Discussion of the Gaussian distribution

Much has been written about the validity of the Gaussian distribution in the theory of errors. Perhaps the best known comment is that experimenters believe in it because they think it can be proved by mathematics, and mathematicians because they think it has been established by experiment. However, the Gaussian distribution does have some theoretical basis. For example, it may be shown to follow from the assumption that each observation is the result of a large number of independent errors, small but discrete, roughly equal in magnitude, and equally likely to be positive or negative.

The assumption of a Gaussian distribution is related to taking the mean of a set of readings as the best value of the measured quantity. The word 'best' in this context is defined as follows. Suppose the distribution function has the form $f(x-X)$, where as usual X is the true value of the quantity. Let ε be the smallest quantity to which the measuring instrument is sensitive. (We suppose ε to be small - its actual value does not affect the argument.) The probability that we shall obtain the values $x_1, x_2, \ldots, x_n$ when we make n measurements is

$$f(x_1-X)f(x_2-X)\ldots f(x_n-X)\varepsilon^n. \tag{3.44}$$

The best value of X is defined to be the one which when inserted into (3.44) makes the quantity a maximum, i.e. it is the value which maximizes the probability of getting the particular set of measured values. Now it can readily be proved that if $f(x-X)$ is a Gaussian, the best value of X is the mean of x_1 to x_n, and conversely that if the best value of X is the mean, then the distribution function is a Gaussian (Whittaker and Robinson 1944, p. 218.)

The Gaussian is the only distribution we shall use, but this should not be taken to imply that all experimental distributions in physics are of this form. Phenomena in which a random process gives rise to discrete measured values - for example, the counting of particles in atomic and nuclear physics - follow the Poisson distribution. This is discussed in Appendix D.

The results we have derived using the Gaussian distribution are in fact rather insensitive to the precise form of the distribution. This applies to the results in sections 3.6, 3.7 and 3.8. We have seen - section 3.7 - that, quite apart from the question of the form of the distribution, the values obtained for the errors are, in the majority of cases, fairly crude estimates. Their uncertainties completely swamp effects due to the distribution being slightly non-Gaussian.

The main thing is to have some distribution which is (a) reasonable and (b) easy to handle algebraically. In most cases the Gaussian distribution fulfils both conditions very nicely. So, unless the measurements provide clear evidence to the contrary, we assume that the distribution is Gaussian and use the formulae based on it. The one *common* instance of non-Gaussian distribution is when the measurements are discrete, being readings of an instrument to the nearest scale division. This situation is discussed in chapter 5.

Summary of symbols, nomenclature, and important formulae

A. *Set of n measurements*

Quantities that are known

measured values	$x_1, x_2, \ldots, x_n$
mean	$\bar{x} = \frac{1}{n}\sum x_i$
residual for ith reading	$d_i = x_i - \bar{x}$
standard deviation of sample	$s = \left(\frac{1}{n}\sum d_i^2\right)^{\frac{1}{2}}$

Quantities that are not known

true value	X
error in ith reading	$e_i = x_i - X$
error in mean	$E = \bar{x} - X$

B. *Distributions*

Distribution of single measurements

standard error $\sigma = \langle e^2 \rangle^{\frac{1}{2}}$

Distribution of means of n measurements

standard error $\sigma_{\mathrm{m}} = \langle E^2 \rangle^{\frac{1}{2}}$

$\langle\ \rangle$ denotes the average over the distribution.

C. *Important relations*

$$\sigma_{\mathrm{m}} = \frac{\sigma}{\sqrt{n}}$$

$$\sigma^2 = \frac{n}{n-1}\langle s^2 \rangle$$

$$\sigma_{\mathrm{m}}^2 = \frac{1}{n-1}\langle s^2 \rangle$$

D. *Formulae for estimating σ and σ_m*

Standard method

$$\sigma \approx \left[\frac{\sum d_i^2}{n-1}\right]^{\frac{1}{2}} = \left[\frac{\sum x_i^2 - \frac{1}{n}(\sum x_i)^2}{n-1}\right]^{\frac{1}{2}}$$

$$\sigma_{\mathrm{m}} \approx \left[\frac{\sum d_i^2}{n(n-1)}\right]^{\frac{1}{2}} = \left[\frac{\sum x_i^2 - \frac{1}{n}(\sum x_i)^2}{n(n-1)}\right]^{\frac{1}{2}}$$

Range method (suitable for $2 \leqslant n \leqslant 12$)

r = difference between largest and smallest reading in set

$$\sigma \approx \frac{r}{\sqrt{n}} \qquad \sigma_{\mathrm{m}} \approx \frac{r}{n}$$

E. Gaussian distribution

$$f(x) = \frac{1}{\sqrt{(2\pi)}} \frac{1}{\sigma} \exp\left[-(x-X)^2/2\sigma^2\right].$$

Put $X = 0$. The fraction of readings between x and $x + \mathrm{d}x$ is $f(z)\,\mathrm{d}z$, where

$$f(z) = \frac{1}{\sqrt{(2\pi)}} \exp(-z^2/2), \qquad z = \frac{x}{\sigma}.$$

The fraction of readings between $-x$ and $+x$ is

$$\phi(z) = \sqrt{\left(\frac{2}{\pi}\right)} \int_0^z \exp(-t^2/2)\,\mathrm{d}t.$$

$f(z)$ and $\phi(z)$ are tabulated in Appendix A.

Exercises

3.1 A group of students measure g, the acceleration due to gravity, with a compound pendulum and obtain the following values in units of m s^{-2}:

9·81, 9·79, 9·84, 9·81, 9·75, 9·79, 9·83.

Set out the values as in Table 3.2, and calculate the mean and the residuals. Hence estimate σ by the standard method. Give the best estimate of g, together with its error, for the group.

3.2 In an undergraduate practical class in the Cavendish Laboratory there was an experiment, originally devised by Searle, to measure the Young modulus E for steel by applying a known load to a rod and measuring the deflection by an optical method based on Newton's rings. Although ingenious and capable of considerable precision in the hands of a skilled experimenter, such as Searle himself, the results obtained by the students were found to have a considerable scatter. The experiment was therefore replaced by one in which a horizontal steel beam was supported near its ends, and the deflection when a known load was applied at the centre was measured directly by a dial indicator.

The values obtained for E by the last 10 students who did the Newton's rings experiment and by the first 10 who did the dial indicator experiment are given below. The values are in units of 10^{11} N m^{-2}.

Newton's rings experiment 1·90, 2·28, 1·74, 2·27, 1·67, 2·01, 1·60, 2·18, 2·18, 2·00.

Dial indicator experiment 2·01, 2·05, 2·03, 2·07, 2·04, 2·02, 2·09, 2·09, 2·04, 2·03.

For each set of values, calculate the mean value of E, and estimate the standard error in the mean, by the two methods given in this chapter. Do the results indicate any systematic difference in the two experimental methods?

3.3 The thermal conductivity of copper at 0 °C is

$$k = 385{\cdot}0 \text{ W m}^{-1} \text{ K}^{-1}.$$

A large number of measurements of k, free from systematic error, form a Gaussian distribution with standard error

$$\sigma = 15{\cdot}0 \text{ W m}^{-1} \text{ K}^{-1}.$$

What is the probability that a single measurement lies in the range
(a) 385·0 to 385·1, (b) 400·0 to 400·1, (c) 415·0 to 415·1,
(d) 370·0 to 400·0, (e) 355·0 to 415·0, (f) 340·0 to 430·0 W m^{-1} K^{-1}?

4

Further topics in statistical theory

4.1 The treatment of functions

In most experiments we do not measure the final quantity Z directly. Instead we measure certain primary quantities A, B, C, etc. and calculate Z, which must be a known function of the primary quantities. For example, we might measure the density d of the material of a rectangular block by measuring the mass M and the dimensions l_x, l_y, l_z of the block. The functional relation between the quantity we require, d, and the primary quantities M, l_x, l_y, l_z is

$$d = \frac{M}{l_x l_y l_z}. \tag{4.1}$$

Suppose that each primary quantity has been measured several times. Then, in the case of A, we have a best value $\bar{A}$, the mean of the measured values, and an estimate of its standard error ΔA. (The latter is the σ_{m} of the previous chapter.) Similarly we have $\bar{B}$ and an estimate of ΔB. We assume that the measurements of the primary quantities are independent and therefore that the errors in them are also independent. By this we mean, for example, that if the value of $\bar{A}$ happens to be high, the value of $\bar{B}$ still has an equal probability of being high or low. From the values $A = \bar{A}$, $B = \bar{B}$, etc., the best value of Z may be calculated. The problem we are going to consider is how to calculate the standard error ΔZ in Z from the standard errors ΔA, ΔB, etc. Although we are restricting the discussion to independent measurements, there are occasional situations where the assumption is not valid. The way ΔZ is calculated in such cases depends on the way the primary errors are related; no general rule can be given. Exercise 4.2 provides an example of related errors.

(*a*) *Functions of one variable.* We consider first the case where Z is a function of only one variable A, for example

$$Z = A^2 \quad \text{or} \quad Z = \ln A.$$

We write this in general as

$$Z = Z(A). \tag{4.2}$$

(The symbol A is used both for the name of the primary quantity and for its value.)

If the true value of the primary quantity is A_0, the true value of Z is

$$Z_0 = Z(A_0) \tag{4.3}$$

- see Fig. 4.1. The error in a given value A is

$$E_A = A - A_0, \tag{4.4}$$

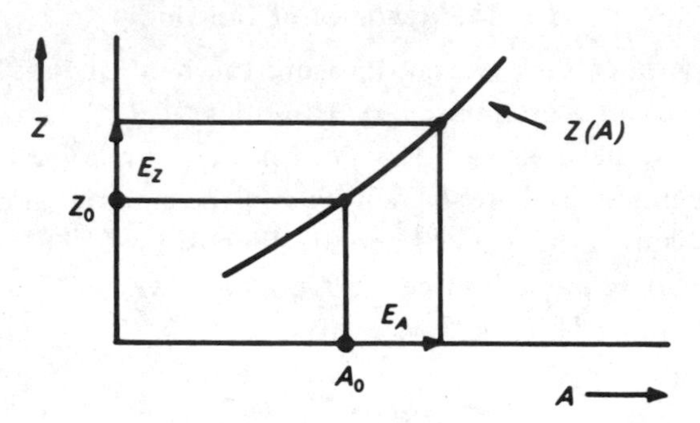

Fig. 4.1. Error E_z in Z due to error E_A in A.

and this gives rise to an error E_Z in Z, where

$$E_Z = Z(A_0 + E_A) - Z(A_0) \tag{4.5}$$

$$\approx \frac{\mathrm{d}Z}{\mathrm{d}A} E_A. \tag{4.6}$$

The derivative $\mathrm{d}Z/\mathrm{d}A$ is evaluated at $A = A_0$. The approximation in (4.6) is equivalent to the assumption that the error in A is sufficiently small for $Z(A)$ to be represented by a straight line over the range of the measured values of A. The error in Z is therefore proportional to the error in A, the constant of proportionality being

$$c_A = \left(\frac{\mathrm{d}Z}{\mathrm{d}A}\right)_{A=A_0}. \tag{4.7}$$

We now allow A to vary according to the distribution of which $\bar{A}$ is a member and take the root-mean-square of (4.6). This gives the result

$$\Delta Z = c_A \Delta A. \tag{4.8}$$

An important special case is $Z = A^n$, for which $c_A = nA^{n-1}$. Then

$$\frac{\Delta Z}{Z} = n\frac{\Delta A}{A}, \tag{4.9}$$

i.e. the fractional standard error in Z is n times that in A. We have already used this result (p. 26) for the case $n=\frac{1}{2}$.

(b) Functions of several variables. We next consider the case where Z is a function of two variables A and B,

$$Z=Z(A, B). \tag{4.10}$$

The errors in A and B are

$$E_A=A-A_0, \qquad E_B=B-B_0, \tag{4.11}$$

where A_0 and B_0 are the true values of A and B. As before we assume that Z is approximately a linear function of A and B in the range over which the measured values vary. Then the error in Z is

$$E_Z=c_A E_A+c_B E_B, \tag{4.12}$$

where the coefficients c_A and c_B are given by

$$c_A=\frac{\partial Z}{\partial A}, \qquad c_B=\frac{\partial Z}{\partial B}. \tag{4.13}$$

The partial derivatives are evaluated at $A=A_0$, $B=B_0$.

From (4.12)

$$E_Z^2=c_A^2 E_A^2+c_B^2 E_B^2+2c_A c_B E_A E_B. \tag{4.14}$$

We take the average of this equation for pairs of values of A and B taken from their respective distributions. Since A and B are assumed independent, the average value of $E_A E_B$ is zero. By definition

$$(\Delta Z)^2=\langle E_Z^2\rangle, \qquad (\Delta A)^2=\langle E_A^2\rangle, \qquad (\Delta B)^2=\langle E_B^2\rangle. \tag{4.15}$$

Therefore

$$(\Delta Z)^2=c_A^2(\Delta A)^2+c_B^2(\Delta B)^2. \tag{4.16}$$

We can now state the general rule. Let Z be a known function of A, B, C, ... Let the standard error in A be ΔA and so on. Then the standard error ΔZ in Z is given by

$$(\Delta Z)^2=(\Delta Z_A)^2+(\Delta Z_B)^2+(\Delta Z_C)^2+\ldots, \tag{4.17}$$

where

$$\Delta Z_A=\left(\frac{\partial Z}{\partial A}\right)\Delta A \quad \text{and so on.} \tag{4.18}$$

The expressions for ΔZ for some common relations between Z and A, B are given in Table 4.1.

Table 4.1. *Combination of errors*

Relation between Z and A, B	Relation between standard errors	
$\left.\begin{array}{l} Z=A+B \\ Z=A-B \end{array}\right\}$	$(\Delta Z)^2=(\Delta A)^2+(\Delta B)^2$	(i)
$\left.\begin{array}{l} Z=AB \\ Z=A/B \end{array}\right\}$	$\left(\frac{\Delta Z}{Z}\right)^2=\left(\frac{\Delta A}{A}\right)^2+\left(\frac{\Delta B}{B}\right)^2$	(ii)
$Z=A^n$	$\frac{\Delta Z}{Z}=n\frac{\Delta A}{A}$	(iii)
$Z=\ln A$	$\Delta Z=\frac{\Delta A}{A}$	(iv)
$Z=\exp A$	$\frac{\Delta Z}{Z}=\Delta A$	(v)

4.2 The straight line

In an experiment it is often the case that one quantity y is a function of another quantity x, and measurements are made of pairs of values of x and y. The values are then plotted on a graph and we try to find a curve corresponding to some algebraic function $y=y(x)$ which passes as closely as possible through the points. We shall only consider the case where the function is the straight line

$$y=mx+c. \tag{4.19}$$

The problem is to calculate the values of the parameters m and c for the best straight line through the points.

The straight-line relation covers a great range of physical situations. In fact we usually try to plot the graph so that the expected relationship *is* a straight line. For example, if we expect the refractive index μ of a certain glass to be related to the wavelength λ of the light by the equation

$$\mu=a+b/\lambda^2, \tag{4.20}$$

we plot μ against $1/\lambda^2$.

We give two methods for calculating the best, i.e. most probable, line through a set of points.

(a) The method of least squares. This is the standard statistical method. Suppose there are n pairs of measurements (x_1, y_1), (x_2, y_2), . . . , (x_n, y_n)

- Fig. 4.2. Assume that the errors are entirely in the y values.* For a given pair of values for m and c, the deviation of the ith reading is

$$y_i - mx_i - c. \tag{4.21}$$

The best values of m and c are taken to be those for which

$$S = \sum(y_i - mx_i - c)^2 \tag{4.22}$$

is a minimum† - hence the name *method of least squares.*

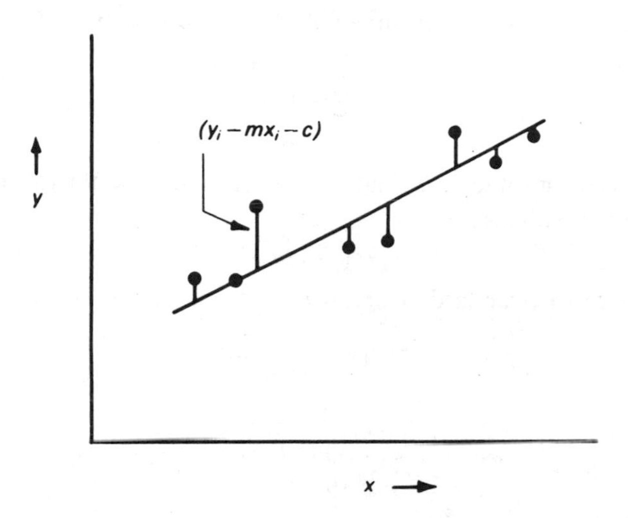

Fig. 4.2. Method of least squares. The best line through the points is taken to be the one for which $\sum(y_i - mx_i - c)^2$ is a minimum.

The principle of minimizing the sum of the squares of the deviations was first suggested by Legendre in 1806. We have already seen that in the case of a single observable the principle gives the mean as the best value.

From (4.22)

$$\frac{\partial S}{\partial m} = -2\sum x_i(y_i - mx_i - c) = 0, \tag{4.23}$$

$$\frac{\partial S}{\partial c} = -2\sum(y_i - mx_i - c) = 0. \tag{4.24}$$

Therefore the required values of m and c are obtained from the two

* The analysis for the case when there are errors in both the x and y variables is much more complicated - see Guest 1961, p. 128 - but the resulting straight line is usually quite close to that given by the present calculation - see exercise 4.4.

† The points are assumed to have equal weights. The case of unequal weights is discussed in the next section.

simultaneous equations

$$m\sum x_i^2 + c\sum x_i = \sum x_i y_i, \tag{4.25}$$

$$m\sum x_i + cn = \sum y_i. \tag{4.26}$$

The last equation shows that the best line goes through the point

$$\bar{x} = \frac{1}{n}\sum x_i, \qquad \bar{y} = \frac{1}{n}\sum y_i, \tag{4.27}$$

i.e. through the centre of gravity of all the points. From (4.25) and (4.26)

$$m = \frac{\sum(x_i - \bar{x})y_i}{\sum(x_i - \bar{x})^2}, \tag{4.28}$$

$$c = \bar{y} - m\bar{x}. \tag{4.29}$$

When the best values of m and c are inserted into (4.21), the deviations become the residuals

$$d_i = y_i - mx_i - c. \tag{4.30}$$

Estimates of the standard errors in m and c are given by

$$(\Delta m)^2 \approx \frac{1}{D}\frac{\sum d_i^2}{n-2}, \tag{4.31}$$

$$(\Delta c)^2 \approx \left(\frac{1}{n} + \frac{\bar{x}^2}{D}\right)\frac{\sum d_i^2}{n-2}, \tag{4.32}$$

$$D = \sum(x_i - \bar{x})^2. \tag{4.33}$$

These results are proved in Appendix E.

If we require the best line that passes through the origin, the value of m is given by (4.25) with $c = 0$:

$$m = \frac{\sum x_i y_i}{\sum x_i^2}. \tag{4.34}$$

An estimate of its standard error is given by

$$(\Delta m)^2 \approx \frac{1}{\sum x_i^2}\frac{\sum d_i^2}{n-1}. \tag{4.35}$$

(b) Programmable calculator or computer. A programmable calculator or computer may be used to calculate $m, c, \Delta m, \Delta c$. The expressions for these quantities in equations (4.28)–(4.33) are not suitable for this purpose as they stand, because the values of $\bar{x}$, $\bar{y}$, m and c (and therefore of d_i) are not known until all the numbers are fed in. However, just as in the calculation of σ (section 3.4 c), the expressions may be recast into a form that avoids this difficulty. Put

$$E = \sum(x_i - \bar{x})(y_i - \bar{y}) = \sum(x_i - \bar{x})y_i, \tag{4.36}$$

$$F = \sum(y_i - \bar{y})^2. \tag{4.37}$$

From (4.28)

$$m = \frac{E}{D}. \tag{4.38}$$

Now

$$\begin{aligned} \sum d_i^2 &= \sum (y_i - mx_i - c)^2 \\ &= \sum [(y_i - \bar{y}) - m(x_i - \bar{x})]^2 \\ &= \sum (y_i - \bar{y})^2 - 2m \sum (x_i - \bar{x})(y_i - \bar{y}) + m^2 \sum (x_i - \bar{x})^2 \\ &= F - 2mE + m^2 D \\ &= F - \frac{E^2}{D}. \end{aligned} \tag{4.39}$$

From (4.31), (4.32) and (4.39)

$$(\Delta m)^2 \approx \frac{1}{n-2} \frac{DF - E^2}{D^2}, \tag{4.40}$$

$$(\Delta c)^2 \approx \frac{1}{n-2} \left(\frac{D}{n} + \bar{x}^2 \right) \frac{DF - E^2}{D^2}. \tag{4.41}$$

Equations (4.33), (4.36) and (4.37) may be written in the form

$$D = \sum x_i^2 - \frac{1}{n} (\sum x_i)^2, \tag{4.42}$$

$$E = \sum x_i y_i - \frac{1}{n} \sum x_i \sum y_i, \tag{4.43}$$

$$F = \sum y_i^2 - \frac{1}{n} (\sum y_i)^2. \tag{4.44}$$

As the pairs of numbers x_i, y_i are fed in, the calculator or computer accumulates the values of $\sum x_i$, $\sum y_i$, $\sum x_i^2$, $\sum x_i y_i$ and $\sum y_i^2$, and evaluates D, E and F from (4.42)-(4.44). The quantities m, c, Δm and Δc are then obtained from (4.38), (4.29), (4.40) and (4.41).

(c) Points in pairs. In the absence of a programmable calculator or a computer, the least-squares calculation is laborious. The following method provides a simple alternative that is often adequate for the purpose. It is particularly useful when the x values are equally spaced.

In order to illustrate the method, suppose that we have 8 points that lie approximately on a straight line, and we require the best value of the slope m and the error in it. Let the points be numbered in order from 1 to 8 - see Fig. 4.3. Consider points 1 and 5; they determine a straight line and hence a value for the slope. Pairing the points in this way we obtain four values for the slope. We take their mean $\bar{m}$ as the best value of m and find its standard error in the usual way.

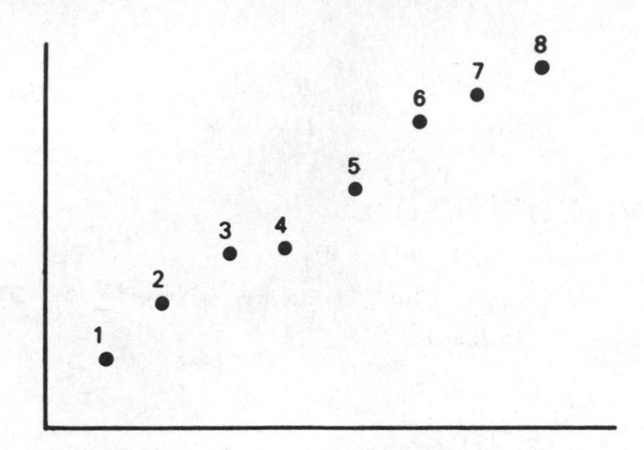

Fig. 4.3. Simple method of estimating slope of best line. Each pair of points 1–5, 2–6, etc. gives a value of the slope. The mean is taken as the best value.

The method will give a reasonable result only if the quantities $(x_5 - x_1)$, $(x_6 - x_2)$, $(x_7 - x_3)$, $(x_8 - x_4)$ are roughly equal. Otherwise, the four values of the slope do not have equal weight.

The best line given by this method is the one with slope $\bar{m}$ that passes through the point $\bar{x}$, $\bar{y}$. (We have already seen that the line given by the method of least squares passes through this point.) However, the method is mainly used when only the slope is required.

4.3 Weighting of results

Suppose we measure a quantity a certain number of times, say 10, and obtain the values $x_1, x_2, \ldots, x_{10}$. Suppose further that we divide the measurements into two sets and calculate the mean of each. For example, we might make 7 of the measurements in the morning and calculate their mean

$$z_1 = \tfrac{1}{7}(x_1 + x_2 + \ldots + x_7). \tag{4.45}$$

Then we might make the other 3 in the afternoon and calculate their mean

$$z_2 = \tfrac{1}{3}(x_8 + x_9 + x_{10}). \tag{4.46}$$

The best value from all 10 measurements is

$$\bar{z} = \tfrac{1}{10}(x_1 + x_2 + \ldots + x_{10}), \tag{4.47}$$

and obviously it is not given by taking the simple mean of z_1 and z_2. If we wish to calculate it from these two quantities, it is given by

$$\bar{z} = \frac{7z_1 + 3z_2}{10}. \tag{4.48}$$

The numbers 7 and 3 are termed the *weights* or *relative weights* of the quantities z_1 and z_2.

In general, if we have a set of N values $z_1, z_2, \ldots, z_N$ with relative weights $w_1, w_2, \ldots, w_N$, then the best value of the quantity is

$$\bar{z} = \frac{\sum w_i z_i}{\sum w_i}. \tag{4.49}$$

If all the ws are multiplied by a constant, the value of $\bar{z}$ is unchanged, so it is only the ratios of the ws that matter.

Suppose now that we have N measurements of the quantity z, each measurement having its own standard error, i.e. we have

$$z_1 \pm \Delta z_1,\ z_2 \pm \Delta z_2, \ldots, z_N \pm \Delta z_N.$$

What weight should we give to each z_i in order to obtain the best value of z from all the measurements? The answer is provided by the simple example at the beginning of the section. We saw that if z_i is the mean of n_i original values, then its weight w_i is proportional to n_i. This assumes that all the original values have the same weight, that is, that they all come from the same distribution characterized by a certain standard error σ. We therefore *imagine* each z_i in the above set is the mean of n_i original values taken from a distribution of standard error σ, and give it weight n_i.

We do not know the value of σ; in fact we can choose it quite arbitrarily but, having fixed on a value, we use the result

$$\Delta z_i = \frac{\sigma}{\sqrt{n_i}} \tag{4.50}$$

to obtain n_i. So

$$w_i = n_i = \frac{\sigma^2}{(\Delta z_i)^2}. \tag{4.51}$$

The standard error in $\bar{z}$ is $\sigma/(\sum n_i)^{\frac{1}{2}}$.

From (4.49) and (4.51) the best value of $\bar{z}$ and its standard error are

$$\frac{\sum(1/\Delta z_i)^2 z_i}{\sum(1/\Delta z_i)^2} \pm \frac{1}{[\sum(1/\Delta z_i)^2]^{\frac{1}{2}}}. \tag{4.52}$$

Both these expressions are independent of the value of σ as of course they must be.

In section 4.2 we gave the method of least squares for finding the best straight line through a set of points of equal weight. The generalization to the case of unequal weights is readily made. If w_i is the weight of the pair of values x_i, y_i, then it is necessary to minimize the quantity

$$S_w = \sum w_i (y_i - m x_i - c)^2. \tag{4.53}$$

The equations for m and c become

$$m\sum w_i x_i^2 + c\sum w_i x_i = \sum w_i x_i y_i, \tag{4.54}$$

$$m\sum w_i x_i + c\sum w_i = \sum w_i y_i. \tag{4.55}$$

The expressions for m and c and their standard errors are given in the summary that follows.

Summary of equations for the best straight line by the method of least squares

n points x_i, y_i

Equal weights

General line $y = mx + c$

$$m = \frac{E}{D} \qquad c = \bar{y} - m\bar{x}$$

$$(\Delta m)^2 \approx \frac{1}{n-2}\frac{\sum d_i^2}{D} = \frac{1}{n-2}\frac{DF-E^2}{D^2}$$

$$(\Delta c)^2 \approx \frac{1}{n-2}\left(\frac{D}{n}+\bar{x}^2\right)\frac{\sum d_i^2}{D} = \frac{1}{n-2}\left(\frac{D}{n}+\bar{x}^2\right)\frac{DF-E^2}{D^2}$$

$$D = \sum x_i^2 - \frac{1}{n}(\sum x_i)^2$$

$$E = \sum x_i y_i - \frac{1}{n}\sum x_i \sum y_i$$

$$F = \sum y_i^2 - \frac{1}{n}(\sum y_i)^2$$

$$\bar{x} = \frac{1}{n}\sum x_i \qquad \bar{y} = \frac{1}{n}\sum y_i$$

$$d_i = y_i - mx_i - c$$

Line through origin $y = mx$

$$m = \frac{\sum x_i y_i}{\sum x_i^2}$$

$$(\Delta m)^2 \approx \frac{1}{n-1}\frac{\sum d_i^2}{\sum x_i^2} = \frac{1}{n-1}\frac{\sum x_i^2 \sum y_i^2 - (\sum x_i y_i)^2}{(\sum x_i^2)^2}$$

$$d_i = y_i - mx_i$$

Unequal weights

General line $y = mx + c$

$$m = \frac{E}{D} \qquad c = \bar{y} - m\bar{x}$$

$$(\Delta m)^2 \approx \frac{1}{n-2}\frac{\sum w_i d_i^2}{D} = \frac{1}{n-2}\frac{DF - E^2}{D^2}$$

$$(\Delta c)^2 \approx \frac{1}{n-2}\left(\frac{D}{\sum w_i} + \bar{x}^2\right)\frac{\sum d_i^2}{D} = \frac{1}{n-2}\left(\frac{D}{\sum w_i} + \bar{x}^2\right)\frac{DF - E^2}{D^2}$$

$$D = \sum w_i x_i^2 - \frac{1}{\sum w_i}\left(\sum w_i x_i\right)^2$$

$$E = \sum w_i x_i y_i - \frac{1}{\sum w_i}\sum w_i x_i \sum w_i y_i$$

$$F = \sum w_i y_i^2 - \frac{1}{\sum w_i}\left(\sum w_i y_i\right)^2$$

$$\bar{x} = \frac{\sum w_i x_i}{\sum w_i} \qquad \bar{y} = \frac{\sum w_i y_i}{\sum w_i}$$

$$d_i = y_i - mx_i - c$$

Line through origin $y = mx$

$$m = \frac{\sum w_i x_i y_i}{\sum w_i x_i^2}$$

$$(\Delta m)^2 \approx \frac{1}{n-1}\frac{\sum w_i d_i^2}{\sum w_i x_i^2} = \frac{1}{n-1}\frac{\sum w_i x_i^2 \sum w_i y_i^2 - \left(\sum w_i x_i y_i\right)^2}{\left(\sum w_i x_i^2\right)^2}$$

$$d_i = y_i - mx_i$$

Exercises

4.1 In the following examples, Z is a given function of the independently measured quantities $A, B, \ldots$ Calculate the value of Z and its standard error ΔZ from the given values of $A \pm \Delta A$, $B \pm \Delta B, \ldots$

(a) $Z = A^2$, $\quad A = 25 \pm 1$.

(b) $Z = A - 2B$, $\quad A = 100 \pm 3$, $B = 45 \pm 2$.

(c) $Z = \dfrac{A}{B}(C^2 + D^{\frac{3}{2}})$, $\quad A = 0{\cdot}100 \pm 0{\cdot}003$, $B = 1{\cdot}00 \pm 0{\cdot}05$, $C = 50{\cdot}0 \pm 0{\cdot}5$, $D = 100 \pm 8$.

(d) $Z = A \ln B$, $A = 10{\cdot}00 \pm 0{\cdot}06$,
$B = 100 \pm 2$.

(e) $Z = 1 - \frac{1}{A}$, $A = 50 \pm 2$.

4.2 The volume V of a rectangular block is determined by measuring the lengths l_x, l_y, l_z of its sides. From the scatter of the measurements a standard error of 0·01% is assigned to each dimension. What is the standard error in V (a) if the scatter is due to errors in setting and reading the measuring instrument and (b) if it is due to temperature fluctuations?

4.3 A weight W is suspended from the centre of a steel bar which is supported at its ends, and the deflection at the centre is measured by means of a dial height-indicator whose readings are denoted by y. The following values are obtained:

W/kg	y/μm
0	1642
$\frac{1}{2}$	1483
1	1300
$1\frac{1}{2}$	1140
2	948
$2\frac{1}{2}$	781
3	590
$3\frac{1}{2}$	426
4	263
$4\frac{1}{2}$	77

(a) Plot the points on a graph and draw the best line by eye. Make an intelligent guess of the standard error in the slope by placing a transparent rule along the points and seeing what might be reasonable limits for the line.
(b) Calculate the best value of the slope and its standard error by the method of least squares, and compare the results with your estimates in (a).
(c) Calculate the best value of the slope and its standard error by the method of points in pairs, and draw the line with this slope through the point $\bar{x}, \bar{y}$. Compare these results with those of (b).

4.4 The zener diode is a semiconductor device with the property that its resistance drops suddenly to almost zero when the reverse bias voltage exceeds a critical value V_z, which depends on the temperature T of the diode. The value of V_z is of the order of volts, but the temperature coefficient $\mathrm{d}V_z/\mathrm{d}T$ is only a few millivolts per °C in the temperature range 20–80 °C. Therefore, to measure $\mathrm{d}V_z/\mathrm{d}T$ precisely, a constant reference voltage is subtracted from V_z, and the resulting voltage V is measured directly on a digital multimeter. The following values are obtained for a particular zener diode:

$T/°C$	V/mV	$T/°C$	V/mV
24·0	72·5	50·0	139
30·0	93	56·2	156·5
37·6	107	61·0	171
40·0	116	64·6	178
44·1	127	73·0	198·5

Treat the data in the same way as in parts (a), (b), and (c) of exercise 4.3, assuming in part (b) that the temperature measurements are free from error. This assumption is probably not correct, so repeat the least squares calculation, assuming that the voltage measurements are free from error, and compare the two values of dV_z/dT.

4.5 The results of the 6 most precise measurements of the mass of the charged π meson given in Wohl *et al.* 1984 are

	Mass of $\pi^{\pm}$/keV	
Year	Value	Standard error
1973	139 569	8
1976	139 571	10
1976	139 568·6	2·0
1976	139 566·7	2·4
1979	139 565·8	1·8
1980	139 567·5	0·9

Calculate the weighted mean and its standard error. (The mass m of a nuclear particle is usually given, as here, in terms of its energy equivalent mc^2.)

5

Common sense in errors

5.1 Error calculations in practice

We are now in a position to estimate the standard errors for a large class of experiments. Let us briefly recapitulate. The final quantity Z is a function of the primary quantities $A, B, C, \ldots$ which are either measured directly or are the slopes or intercepts of straight lines drawn through points representing directly measured quantities.

If the quantity is measured directly, we take the mean of several values to be the best value and obtain its standard error by one of the methods given in chapter 3. (During the present chapter we shall drop the word 'standard' in 'standard error'. We shall not be considering the *actual* error in a measured quantity, and the word 'error' will refer to the standard error, i.e., the standard deviation of the distribution of which the quantity is a member.) If the quantity is the slope or intercept in a straight line, its value and error are obtained either from the method of least squares or from the method of taking the points in pairs.

The best value of Z is calculated from the best values of the primary quantities, and its error is obtained from their errors by the rules given in Table 4.1, or in general from (4.17) and (4.18).

There are often a large number of primary quantities to be measured, and it might be thought that the calculation of the error in each one and the subsequent calculation of the error in Z would be a laborious process. And with many students it is indeed. They calculate the standard deviation automatically for every set of measurements, and combine all the errors irrespective of their magnitudes according to the formal rules, involving themselves in elaborate calculations and ending up with an error calculated to a meaningless number of decimal places, which is usually wrong by several orders of magnitude due to various arithmetical slips on the way.

To see what is required in practice, let us first remember *why* we estimate errors. It is to provide a measure of the significance of the final result. The use made of the error is seldom based on such precise calculation that we need its value to better than 1 part in 4. Often we are interested in the error to much less precision, perhaps only to within a factor of 2. However, let us take 1 part in 4 as an arbitrary but adequate degree of precision for the final error.

(a) Combining errors. If we look at the equation for combining errors (4.17), we see that, owing to the procedure of squaring the terms, one error is often negligible compared with another. Consider the case

$$Z = A + B, \tag{5.1}$$

and let $\Delta A = 2$ and $\Delta B = 1$. From Table 4.1, (i)

$$\Delta Z = (2^2 + 1^2)^{\frac{1}{2}} = 2{\cdot}24. \tag{5.2}$$

So even though ΔB is as much as one-half of ΔA, ignoring ΔB altogether and putting $\Delta Z \approx \Delta A = 2$ makes a difference of only about 1 part in 8 in the final error. If Z is the sum of several quantities, ignoring errors that are one-half of the largest error may be rather drastic, but we shall nearly always be justified in ignoring an error less than one-third of the largest error.

We may notice also the situation when the quantities themselves differ greatly in magnitude. For example, suppose in (5.1) that B is some small correction term and we have values

$$A = 100 \pm 6$$
$$B = \quad 5 \pm ?$$

The error in B will be negligible unless it is as much as 3, but such an error amounts to 60% of B; so the quantity will have to be measured very roughly indeed if its error is to contribute.

In the case of multiplication and division - Table 4.1, (ii) - we add the squares, not of the errors themselves, but of the fractional errors. So in this case, all fractional errors less than about one-third of the largest fractional error may be neglected.

(b) Contributing and non-contributing errors. With these considerations in mind let us go back to the estimation of the errors in the primary quantities. We may call a quantity *contributing* or *non-contributing* according to whether or not its error contributes appreciably to the final error. A quantity may be non-contributing either because it is measured relatively precisely or because it is added to a much larger quantity.

If we suspect that a quantity is non-contributing, it is sufficient to estimate its error very roughly, provided the estimate is on the high side. The reason for this condition is obvious. It ensures that we do not omit the error unjustifiably. If the inflated error is negligible we are quite safe. If not, we must go back to the measurements and work out the error more carefully.

For example, suppose the results of successive weighings of an object are:

$$50{\cdot}3853\ \text{g}$$
$$50{\cdot}3846$$
$$50{\cdot}3847$$
$$50{\cdot}3849.$$

We take the best value of the weight to be

$$50{\cdot}3849 \pm 0{\cdot}0003\ \text{g}.$$

We expect this set of measurements to be much more precise than several others in the particular experiment and we therefore estimate an error simply by inspecting the measurements. The value 0·0003 encompasses 3 out of the 4 individual readings, so it is almost certain to be an overestimate of the error in the mean.

(*c*) *Discrete readings*. Another case where a common-sense estimate of the error should be made is when the readings are in digital form or are taken to the nearest scale division of an instrument, and show little or no spread. Consider the following set of measurements made with a metre rule:

$$325,\ 325,\ 325,\ 325\tfrac{1}{2},\ 325,\ 325\ \text{mm}.$$

The most one cay say is that the measured quantity is $325 \pm \frac{1}{2}$ mm or $325 \pm \frac{1}{4}$ mm.* If a better value of the quantity and its error are required, they will not be obtained by more arithmetic, nor by more measurements of the same kind. Either the scale should be estimated to $\frac{1}{10}$ mm as the measurements are made, or a more precise instrument such as a cathetometer should be used.

(*d*) *Systematic errors*. So far we have confined the discussion to the estimation of random errors. And this is all that is needed in the majority of experiments. Any systematic error that we know about should be corrected and hence eliminated – or at least rendered negligible. Normally we would reduce it to a level small compared with the random errors. So it would be non-contributing and would not enter the error calculation.

* It is not unknown for students to solemnly feed these numbers into their calculators, arriving at the result $325{\cdot}08 \pm 0{\cdot}08$ mm.

The occasional situation when residual systematic errors are not small compared with random errors should be discussed and treated on its merits. One way of proceeding is to try to estimate, for each systematic error, something equivalent to a standard error, that is to say, a quantity such that we think there are about 2 chances in 3 that the true value lies within the quoted range. For example, we might estimate - or make an intelligent guess of - an upper limit, and then divide it by 2. (This may seem rough and ready, but a crude estimate is better than none at all.) All the errors are then combined as though they were random and independent. When this is done, it is good practice to make quite clear how much of the final error is due to the actual random error and how much to the various systematic errors.

(*e*) *The final quoted error*. We may sum up as follows. Systematic errors are eliminated as far as possible. The random errors in contributing quantities are calculated by an appropriate statistical method. Other errors are estimated roughly, the estimates being slightly on the high side. A check - which can often be done mentally - is made that these errors are in fact negligible. The contributing errors are then combined according to the rules of Table 4.1 to give the final quoted error. This quantity represents our best estimate of the standard deviation of the distribution of results that would be obtained if the entire experiment were repeated many times with the same or similar apparatus. It is thus a measure of the overall *reproducibility* of the result.

Some experimenters, having obtained the overall error in the usual way, then proceed to enlarge it by an arbitrary factor to take account of possible, but unknown, sources of systematic error. This is highly undesirable. It is difficult for anyone to make use of these subjective overestimates. You should estimate the error as honestly as you are able and leave it at that. If it subsequently turns out that the 'true' value of the quantity being measured is several times your estimated error away from the value you have obtained, you may or may not be held at fault. But you must not arbitrarily double or treble the error as a kind of safety measure to prevent the situation arising. Quite apart from the confusion caused by the uncertain significance of the final error, the procedure may obscure genuine discrepancies between one experimental result and another, or between theory and experiment.

It is conventional to quote the final error in absolute terms and not as a fraction or percentage. The final value of the quantity being measured and its error should be given to the same number of digits, which should not be more than are meaningful. In general this corresponds to an error of one significant digit, though, if this digit is 1 or 2 a second digit might be given. The fact that we do not want an estimate of the final error more precise than this means that the whole error calculation should be done only to one or at the most two significant digits.

5.2 Complicated functions

The evaluation of quantities of the type $\partial Z/\partial A$ in (4.18) is sometimes quite laborious. As an example consider the measurement of the refractive index μ of a glass prism by measuring the angle A of the prism and the angle D of minimum deviation. The refractive index is obtained from the equation

$$\mu = \frac{\sin \frac{1}{2}(A+D)}{\sin \frac{1}{2}A}. \tag{5.3}$$

The error in μ is given by

$$(\Delta\mu)^2 = (\Delta\mu_A)^2 + (\Delta\mu_D)^2. \tag{5.4}$$

$\Delta\mu_A$ is the error in μ due to the error ΔA in A and is given by

$$\Delta\mu_A = \left(\frac{\partial\mu}{\partial A}\right)\Delta A. \tag{5.5}$$

Similarly for $\Delta\mu_D$.

The expressions for $\partial\mu/\partial A$ and $\partial\mu/\partial D$ are

$$\frac{\partial\mu}{\partial A} = \frac{1}{2} \cdot \frac{\cos \frac{1}{2}(A+D)}{\sin \frac{1}{2}A} - \frac{1}{2}\frac{\sin \frac{1}{2}(A+D)}{\sin \frac{1}{2}A \tan \frac{1}{2}A}, \tag{5.6}$$

$$\frac{\partial\mu}{\partial D} = \frac{1}{2}\frac{\cos \frac{1}{2}(A+D)}{\sin \frac{1}{2}A}, \tag{5.7}$$

These expressions have to be evaluated at $A = \bar{A}$, the measured value of A, and $D = \bar{D}$, the measured value of D. And, provided we do the arithmetic correctly and remember to express ΔA and ΔD in radians, we shall get the right answer for $\Delta\mu_A$ and $\Delta\mu_D$.

However, there is a quicker method. Consider the significance of $\Delta\mu_A$. It is the change in the value of μ when A changes by an amount ΔA, the value of D remaining constant. So it may be obtained by calculating μ from (5.3), first for $A = \bar{A}$, $D = \bar{D}$ and then for $A = \bar{A} + \Delta A$, $D = \bar{D}$. The difference is $\Delta\mu_A$. Similarly $\Delta\mu_D$ is obtained by calculating μ for $A = \bar{A}$, $D = \bar{D} + \Delta D$. All we need are the sine values. We do not have to do any complicated algebra or arithmetic – fruitful sources of mistakes – nor bother to convert ΔA and ΔD into radians. We combine $\Delta\mu_A$ and $\Delta\mu_D$ in the usual way.

The fact that this method of calculating $\Delta\mu_A$ and $\Delta\mu_D$ is much quicker than the more formal method should not lead you to imagine that it is in any way less rigorous or exact. The two methods usually give the same answer, and when they do not, the formal method is not valid. This may be seen from Fig. 5.1, where the results of the two methods are shown for some relation $Z = Z(A)$. The best value of A is $\bar{A}$ and this corresponds to Z_1. The error ΔZ_t obtained by the formal method corresponds to putting the tangent to the curve $Z(A)$ at the point $\bar{A}$, Z_1. The value calculated from the simpler and more direct method is ΔZ_+ in the diagram. We could equally well have calculated ΔZ by taking the value $\bar{A} - \Delta A$, which would have given ΔZ_-.

The curvature of the function $Z(A)$ over the range $\bar{A} + \Delta A$ is not usually as large as that shown in Fig. 5.1; in which case the difference between ΔZ_t, ΔZ_+ and ΔZ_- is negligible. If, however, the curvature of the function is significant, then a single value of the error is misleading. Instead, both ΔZ_+ and ΔZ_- should be calculated and the result quoted as

$$Z = Z_1 \begin{array}{l} + \Delta Z_+ \\ - \Delta Z_- \end{array}.$$

Such refinement is seldom justified. The main point is that to calculate the error in Z due to an error in A, we cannot go wrong if we simply

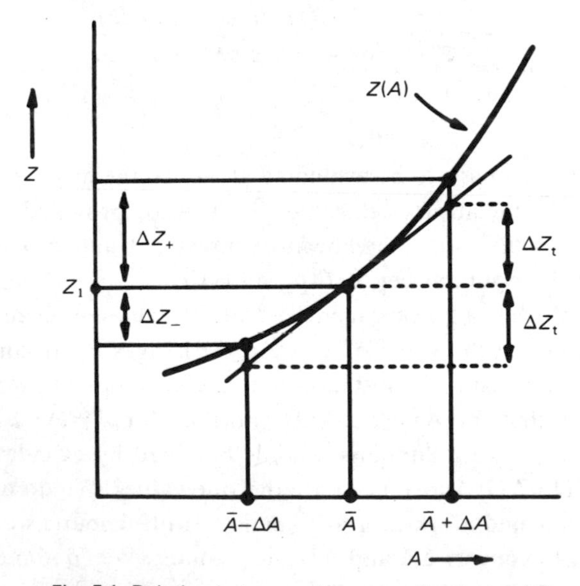

Fig. 5.1. Relation between different estimates of ΔZ.

calculate the values of Z at the values $\bar{A}$ and $\bar{A}+\Delta A$, with the other measured quantities constant. And often this is much quicker than the formal method.

5.3 Errors and experimental procedure

When the final quantity Z is related to two directly measured quantities by a function of the form

$$Z = AB \quad \text{or} \quad A/B,$$

then an error of $x\%$ in A or B gives rise to an error of $x\%$ in Z. So we would try to measure A and B with comparable precision, and this is true whatever the relative magnitudes of A and B. But the situation

$$Z = A + B \quad \text{or} \quad A - B$$

is quite different. Everything depends on the relative magnitudes of A and B. Look at the following example:

$$\begin{aligned} \text{Case I} \qquad\qquad A &= 10\,000 \pm 1, \\ B &= \quad 100 \pm 5, \\ Z = A + B &= 10\,100 \pm 5. \end{aligned}$$

Here A is a large, precisely known quantity. B has been measured to 5%, but the final quantity Z has been found to 0·05%. So we see that it is advantageous to start with a large, precisely known quantity and simply measure a small additional term in order to get the required quantity.

Now consider the following:

$$\begin{aligned} \text{Case II} \qquad\qquad A &= 100 \pm 2, \\ B &= \;\; 96 \pm 2, \\ Z = A - B &= \quad 4 \pm 3. \end{aligned}$$

The two directly measured quantities have been determined to 2%, but the final quantity is only known to 75%. So taking the difference between two nearly equal quantities, each of which is measured independently, is inherently disadvantageous; the final error is greatly magnified. If possible an entirely different method of measuring Z should be found.

In the next two chapters we shall give specific examples of methods devised to take advantage of the Case I situation and others devised to avoid Case II. They provide examples of the way error considerations may have a direct influence on experimental procedure.

We give one hypothetical example here. Consider the following situation. We require to measure the quantity $Z = A/B$. We have made a set of measurements and found

$$A = 1000 \pm 20,$$
$$B = \quad 10 \pm 1.$$

Therefore

$$\frac{\Delta A}{A} = 2\% \quad \text{and} \quad \frac{\Delta B}{B} = 10\%,$$

whence

$$\frac{\Delta Z}{Z} = (2^2 + 10^2)^{\frac{1}{2}} = 10{\cdot}2\%.$$

We have some further time available for measurements and estimate it is sufficient to reduce the error in either A or B by a factor 2. If we devote the time to A, we shall have

$$\frac{\Delta A}{A} = 1\%, \quad \text{which gives} \quad \frac{\Delta Z}{Z} = (1^2 + 10^2)^{\frac{1}{2}} = 10{\cdot}0\%.$$

If we devote it to B, we shall have

$$\frac{\Delta B}{B} = 5\%, \quad \text{which gives} \quad \frac{\Delta Z}{Z} = (2^2 + 5^2)^{\frac{1}{2}} = 5{\cdot}4\%.$$

So in the first case the overall error is barely changed, and in the second case it is reduced by a factor of almost 2. The moral is *always concentrate on quantities that contribute most to the final error.*

In general one should plan the experiment so that in the final result no one quantity contributes an error much greater than the others. In the present example we may suspect that the original measurements, which resulted in $\Delta B/B$ being 5 times greater than $\Delta A/A$, were badly planned, and that more time should have been devoted to measuring B at the expense of A. Of course it is not always the case that additional measurements result in a reduction of the error. Nevertheless, the desirability of reducing the maximum contributing error should always be kept in mind when planning an experiment.

Exercises

5.1 A rectangular brass bar of mass M has dimensions a, b, c. The moment of inertia I about an axis in the centre of the ab face and perpendicular to it is

$$I = \frac{M}{12}(a^2 + b^2).$$

The following measurements are made:

$$M = 135{\cdot}0 \pm 0{\cdot}1 \text{ g},$$
$$a = 80 \pm 1 \text{ mm},$$
$$b = 10 \pm 1 \text{ mm},$$
$$c = 20{\cdot}00 \pm 0{\cdot}01 \text{ mm}.$$

What is the percentage standard error in (a) the density ρ of the material, and (b) the moment of inertia?

5.2 When a torsion wire of radius r and length l is fixed at one end and subjected to a couple of moment C at the other, the angular displacement ϕ is given by

$$\phi = \frac{2lC}{n\pi r^4},$$

where n is the rigidity modulus of the material of the wire. The following values are obtained:

$$\phi/C = 4{\cdot}00 \pm 0{\cdot}12 \text{ rad N}^{-1}\text{ m}^{-1},$$
$$r = 1{\cdot}00 \pm 0{\cdot}02 \text{ mm},$$
$$l = 500 \pm 1 \text{ mm}.$$

Calculate the value of n and its standard error.

5.3 If a narrow collimated beam of monoenergetic γ-rays of intensity I_0 is incident on a thin sheet of material of thickness x, the intensity of the emerging beam is given by

$$I = I_0 \exp(-\mu x),$$

where μ is a quantity known as the linear attenuation coefficient. The following values are obtained for γ-rays of energy 1 MeV incident on lead:

$$I = (0{\cdot}926 \pm 0{\cdot}010) \times 10^{10} \gamma\text{-rays m}^{-2}\text{ s}^{-1},$$
$$I_0 = (2{\cdot}026 \pm 0{\cdot}012) \times 10^{10} \gamma\text{-rays m}^{-2}\text{ s}^{-1},$$
$$x = (10{\cdot}00 \pm 0{\cdot}02) \text{ mm}.$$

Calculate the value of μ and its standard error for γ-rays of this energy in lead.

5.4 Neutrons reflected by a crystal obey Bragg's law $n\lambda = 2d \sin\theta$, where λ is the de Broglie wavelength of the neutrons, d is the spacing between the reflecting planes of atoms in the crystal, θ is the angle between the incident (or reflected) neutrons and the atomic planes, and n is an integer. If n and d are known, the measured value of θ for a beam of monoenergetic neutrons determines λ, and hence the kinetic energy E of the neutrons. If $\theta = 11°18' \pm 9'$, what is the fractional error in E?

5.5 As the temperature varies, the frequency f of a tuning fork is related to its linear dimensions L and the value of the Young modulus E of its material by

$$f \propto \sqrt{(EL)}.$$

When the temperature rises by 10 K, the frequency of a certain fork falls by $(0{\cdot}250 \pm 0{\cdot}002)\%$. For the same temperature rise, the Young modulus of the material falls by $(0{\cdot}520 \pm 0{\cdot}003)\%$. Calculate the value of α, the linear expansivity of the material, given by these experiments. What is its standard error? Is this a good method for measuring the linear expansivity?

PART 2

Experimental methods

6

Some laboratory instruments and methods

6.1 Introduction

In this chapter we shall consider some general principles for making measurements. These are principles which should be borne in mind, first in selecting a particular method and second in getting the most out of it. By the latter we mean making the method as precise or reproducible as possible, and - even more important - avoiding its inherent systematic errors.

We shall illustrate the various points by describing some specific examples of instruments and methods. Though chosen from several branches of physics, they are neither systematic nor exhaustive. The idea is that, having seen how the principles apply in these cases, you will be able to apply them yourselves in other situations. As always there is no substitute for laboratory experience. But experience without thought is a slow and painful way of learning. By concentrating your attention on certain aspects of measurement making we hope to make the experience more profitable.

6.2 Metre rule

We start with almost the simplest measuring device there is - a metre rule. Its advantages are that it is cheap to make and convenient to use. It can give results accurate to about $\frac{1}{5}$ mm. However, to achieve this accuracy certain errors must be avoided.

(a) Parallax error. If there is a gap between the object being measured and the scale, and the line of sight is not at right angles to the scale, the reading obtained is incorrect (Fig. 6.1a). This is known as a *parallax error* and clearly may occur, not only in a metre rule, but in all instruments where a pointer moves over a scale. It may be reduced by having the object or pointer as close to the scale as possible, and also by having a mirror next to the scale as shown in Fig. 6.1b. Aligning the image of the

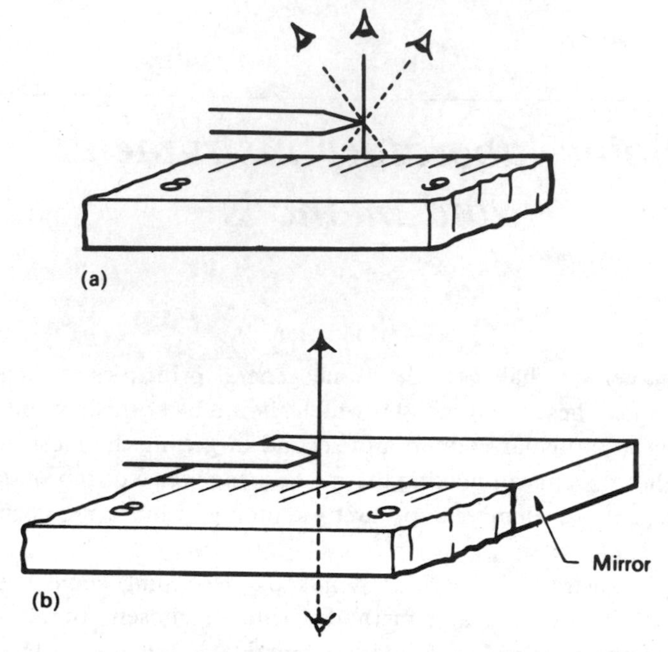

Fig. 6.1. Parallax error. (a) Different positions of the eye result in different readings. (b) A mirror placed beside the scale ensures that the line of sight is at right angles to the scale.

eye with the object ensures that the line of sight is at right angles to the scale.

(b) Zero error. Except for crude measurements it is not good practice to place one end of the rule against one end of the object and take the reading at the other end (Fig. 6.2a). Instead, the object should be placed so that a reading can be taken at both its ends (Fig. 6.2b). This is because the end of the rule may be worn or the zero ruling incorrect in some way. In general the zero position of any instrument should be regarded as suspect. The resulting error can usually be avoided by a simple subtraction technique as here.

(c) Calibration. The scale on the rule may be incorrectly marked. The rule should therefore be checked, or calibrated. This is done by simply laying it beside a more accurate standard rule and noting the readings.

It is important to realize the logic of this. The ordinary metre rule is cheap because it is made of a cheap material - wood - and the scale is

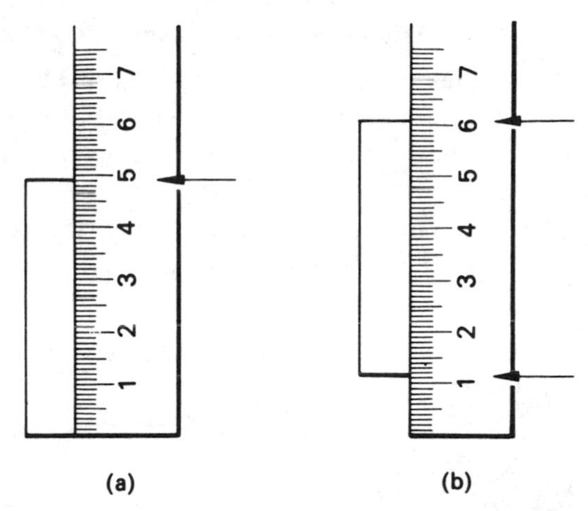

Fig. 6.2. Measuring the length of an object as in (a) is bad practice. It will give a systematic error if the end of the rule is worn. The rule should be placed as in (b) and two readings taken.

engraved without a great deal of care. The two factors go together; it is not worth engraving an accurate scale on a wooden rule whose whole length is liable to change with time.

Suppose we have say 20 experimenters in a laboratory, and they need to measure lengths of about 500 mm to an accuracy of $\frac{1}{10}$ mm. We could supply each one with a steel rule known to be marked to this accuracy. Such a rule is far more expensive than an ordinary wooden one, whose markings are probably only good to about $\frac{1}{2}$ mm over the whole rule. (Plastic rules are much worse than this and are often in error by as much as 1%.) Alternatively, we could supply each person with an ordinary rule and install *one* expensive standard in the laboratory. The cost of this would only be a fraction of the other. But the measurements could still be made to the required accuracy, *provided each experimenter remembers to do the calibration.* This procedure, viz.

many cheap + one expensive standard + calibration

is obviously a sensible one.

If you use only part of the rule during a particular experiment, the comparison with the standard should be done particularly carefully for the part actually used. It would not matter if the rest was incorrectly marked. In practice it is unlikely that one part of a metre rule is ruled less accurately than another. But it is a good general principle when calibrating an instrument to concentrate on the range actually used.

6.3 Micrometer screw gauge

This instrument measures the external dimensions of objects up to the order of 100 mm. A typical micrometer is shown in Fig. 6.3. The spindle has two threads to the millimetre, so that one complete rotation of the thimble T corresponds to 500 μm. The instrument can easily be read to 10 μm, a gain in precision of about 20 over the metre rule. High precision versions of the instrument can be read to 1 μm.

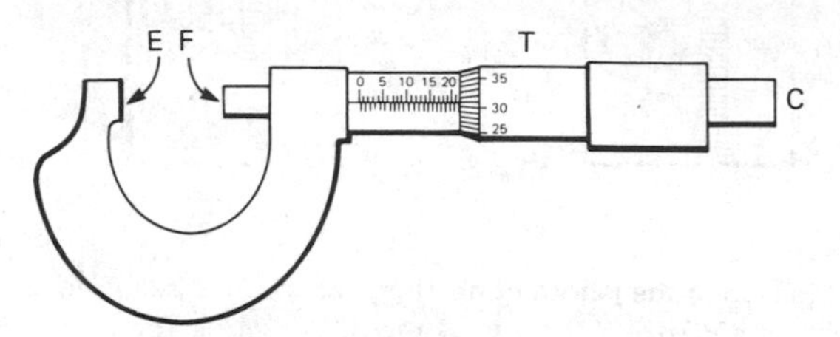

Fig. 6.3. Micrometer screw gauge.

Several points may be noted:

(a) The increase in precision comes from the screw mechanism - a highly effective way of magnifying linear motion.

(b) The personal element involved in the torque applied to the thimble, when the face F is in contact with the object, is eliminated by a ratchet device. The thimble is rotated by applying torque, not to the body itself, but to the end cap C. This transmits the motion only up to a certain standard torque; thereafter it slips, and rotating it further has no effect on the reading. So the final reading corresponds to a standard pressure of the face F against the object.

(c) The instrument is prone to zero error, which should always be checked by taking a reading with the face F right up against E.

(d) Other readings may be checked by means of gauge blocks, which are rectangular blocks of high-grade hardened steel. The end-faces of a block are flat and parallel, and the distance between them, known to about 0·1 μm, is engraved on the block.

6.4 Measurement of length – choice of method

In the last two sections we have considered two methods of measuring length. We are not going to describe any more in detail, but, instead, shall look at the general problem of measuring length.

First we must decide what we mean by length. On the whole we know fairly well what we mean in the case of objects whose velocity is small compared with that of light, and which we can see, or almost see, in the laboratory or near the Earth and Sun – a range in length say of about 10^{-8} to about 10^{11} m. But when we consider objects whose velocity is not small compared with that of light, or try to extend the range downwards to objects as small as atoms and elementary particles, or upwards to the distances that separate us from the stars and galaxies, we have to say what we mean by length and distance – we cannot lay a ruler across a nucleus or extend one to a star. We have to say something like 'if we do such and such an experiment, the result is as though something extends over such and such a distance, or is a certain distance away'. Such *operational* definitions sometimes lead to concepts of length and distance which differ from our usual common-sense ideas on the subject. But the latter have come from seeing and thinking about lengths over a very restricted range. So we need not be too surprised if they do not apply when the range is enormously extended.

The problems of definition and measurement at the extreme ends of the range are outside the scope of this book. We shall only consider the range where our common-sense ideas do apply. However, it does no harm to put the discussion in a more general context and to remind ourselves that behind *all* measurements lies the definition – usually implicit – of the quantity being measured.

Even restricting ourselves to measurements of length of the order of 1 μm to 1 m, we have a range of instruments to choose from. In deciding which one to use we should consider the following:

(a) The nature of the length we wish to measure, for example, whether it is the distance between two marks or between the two end-faces of a rod or bar, or the diameter of a hole or of a rod.
(b) The rough value of the length.
(c) The precision required.

In Table 6.1 we list five instruments which are available in the range we are considering. Vernier calipers are shown in Fig. 6.4. A cathetometer is simply a telescope that can slide on a rigid graduated bar. A travelling microscope is an instrument in which the distance travelled by the microscope, from a position in which one object or mark is in focus on its crosswires to another similar position, is given by the rotation of a screw device that controls the motion. (This instrument is subject to the error known as 'backlash'. Owing to looseness in the moving parts, the reading depends on the direction from which the crosswires are moved into position. The error is avoided if the final setting is always made

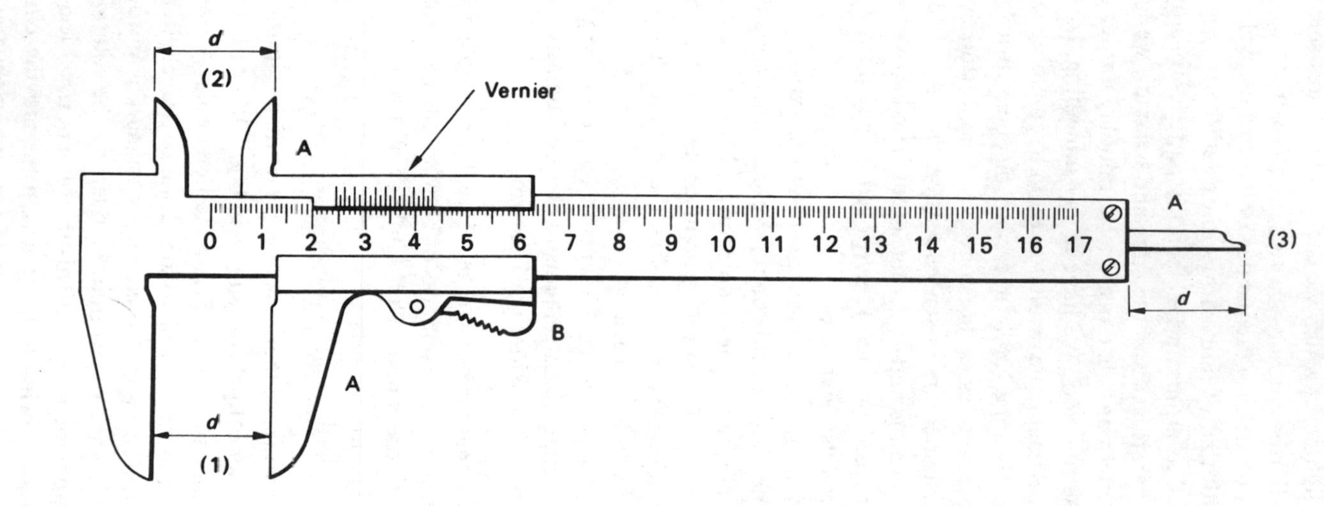

Fig. 6.4. Vernier calipers. The parts marked A form a rigid unit, which is free to move relative to the rest of the instrument when the spring-loaded button B is pressed. The three distances marked d are equal and are read off from the vernier scale. (1) gives the diameter of a rod, (2) the diameter of a hole and (3) the depth of a blind hole.

Table 6.1. *Suitability of some instruments for measuring length*

Instrument	Range/m	Precision/μm	Applicability
metre rule	1	200	general
vernier calipers	0·1	50	most useful for overall dimensions of an object, also for dimensions of a space – width of gap, diameter of hole
cathetometer	1	10	general
micrometer screw gauge	0·1	2	overall dimensions of an object
travelling microscope	0·2	1	general

from the same direction.) The table shows the approximate range and precision, and type of measurement for each instrument.

Of course, having chosen the right instrument we have to make the right set of measurements with it. In theoretical physics we say a cylindrical rod has a diameter of d mm and that is the end of the matter. But in practical physics we have to verify that the rod *is* a cylinder – or, to be more correct, to say within what limits it is. So at one place along the length we should measure the diameter in various directions. Then we should repeat the measurements at various places along the length. As always, the thoroughness of the investigation depends on the purpose of the measurements. Similarly, before giving a distance between the end faces of a bar, we must check the extent to which the faces are parallel.

We mention briefly how the range of lengths and the precision of the measurements may be increased. The latter may be achieved by optical interference experiments. Similar experiments with various forms of waves – light, X-rays, electrons, neutrons – also extend the range in the downward direction. Another way of measuring changes in length is to measure the change in capacitance when one plate of a capacitor moves relative to another (Sydenham 1985). At the upper end of the length range we may measure distances by triangulation. The enormous cosmological distances are deduced – indirectly and with varying degrees of plausibility – from measurements of quantities such as the apparent brightness of stars (Rowan-Robinson 1981).

All is grist to the physicist's mill. We look for any physical phenomenon that depends on the quantity we wish to measure. From the variety of instruments based on these phenomena, we select one of appropriate applicability, range, and precision.

6.5 Measurement of length – temperature effect

In any precise measurement of length we must consider the effects of thermal expansion. This applies first to the object being measured. Suppose, for example, we require the length of a tube at liquid hydrogen temperature (20 K). If we measured its length at room temperature and did not allow for the contraction, we should make a serious error. In this extreme case, to calculate the correction we could not assume that the linear expansivity was independent of temperature, but would have to know its values from 300 K down to 20 K.

Second, we must consider whether thermal expansion is affecting the readings of the instrument used to measure the lengths. And last, if we calibrate the instrument by measuring the length of a standard object such as a gauge block, we must consider whether the length of the latter has departed from its nominal value owing to temperature change.

In Table 6.2 the values of α, the linear expansivity, for a few common substances are given. The values are approximate and correspond to room temperature (293 K). Invar is a steel alloy containing 36% nickel; its chief merit is its very low expansivity at room temperature, and it is often used when this property is required. At higher temperatures its value of α increases. Fused silica is superior as a low-expansion substance; the value of α given in the table applies up to 1300 K. Moreover, it is extremely stable dimensionally. It is therefore often used in precise measurements where the geometry must be very exactly defined and must vary as little as possible with temperature.

Let us look at the actual values in the table. The substances listed are fairly representative. Most solids at room temperature have values of α

Table 6.2. *Linear expansivity α for some common substances*

Substance	$\alpha/10^{-6}\ K^{-1}$
copper	17
brass	19
steel	11
Invar	1
soda glass	9
Pyrex	3
fused silica	0·5
wood – along grain	4
across grain	50

in the range 5 to $25\times10^{-6}\,\mathrm{K}^{-1}$. If we take $10^{-5}\,\mathrm{K}^{-1}$ as a typical value, we see that a change in temperature of 10 K corresponds to a change in length of 1 part in 10^4.

It is very important in practical physics to have an idea of orders of magnitude of this kind. On the one hand, if we do not correct for thermal changes in dimensions in a precise measurement, we may make a serious error. On the other hand, if we were to calculate the possible thermal correction every time we measured a length of any kind, we would obviously waste a lot of time. Common sense tells us this. But what common sense does not tell us is at what stage, as the precision of an experiment is increased, we should start to concern ourselves with the effect. This is where the order of magnitude calculation comes in. It tells us that, for temperature changes of less than 10 K, we need not worry about temperature effects in length, unless we are measuring it to better than 1 part in 10^4.

6.6 The beat method of measuring frequency

(a) The phenomenon of beats. Suppose we have two sinusoidal waves of equal amplitude A but slightly different frequencies f_1 and f_2. We do not specify the physical nature of the waves. They may, for example, be the displacement in space of a mechanical vibrating system or the voltage across a capacitor in an oscillatory circuit. Whatever their nature we can represent the separate displacements by

$$y_1 = A\cos 2\pi f_1 t \quad \text{and} \quad y_2 = A\cos 2\pi f_2 t. \tag{6.1}$$

They are shown in Fig. 6.5a and b. If the two waves act together, the total displacement is

$$\begin{aligned} y = y_1 + y_2 &= A(\cos 2\pi f_1 t + \cos 2\pi f_2 t) \\ &= 2A\cos 2\pi\frac{f_1 - f_2}{2}t\cos 2\pi\frac{f_1+f_2}{2}t. \end{aligned} \tag{6.2}$$

Suppose f_1 and f_2 are fairly close in value, that is,

$$f_1 - f_2 \ll f_1 + f_2. \tag{6.3}$$

Then we may regard the factor $\cos 2\pi[(f_1+f_2)/2]t$ in (6.2) as representing rapid sinusoidal motion, the amplitude of which is varying slowly according to the factor $2A\cos 2\pi[(f_1-f_2)/2]t$. The overall motion is shown in Fig. 6.5c.

Quite apart from (6.2), derived from the previous line by straightforward trigonometry, we can see how the waves in (a) and (b) in Fig. 6.5 add to give the wave in (c). At time P the two waves are in phase, and their

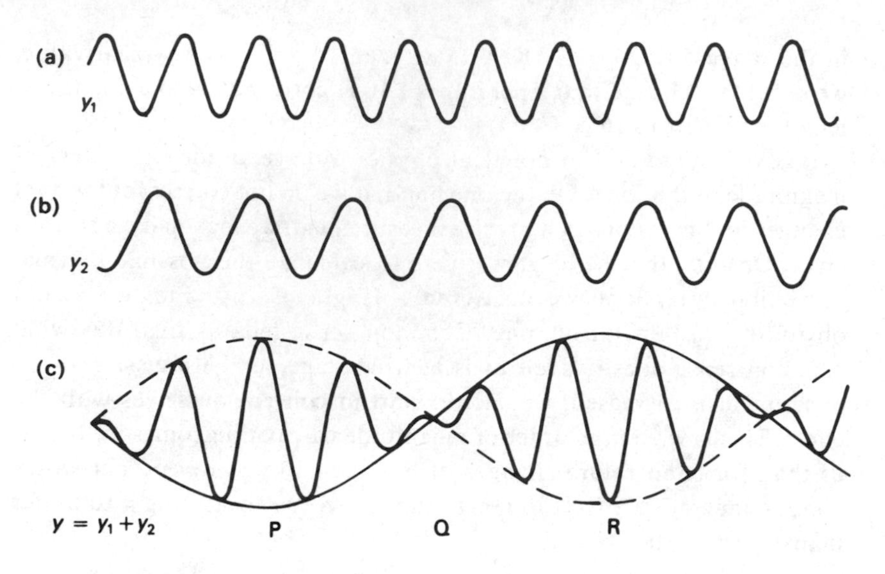

Fig. 6.5. The two waves y_1 and y_2, of slightly different frequency, add to give the resultant shown in (c).

sum is large. As their frequencies are slightly different, they gradually get out of phase until at Q they are exactly 180° out of phase, and their sum is zero. At R they are back in phase again. This phenomenon, the successive swelling up and dying down of the resultant, is known as *beats*, and the frequency of the maxima in the amplitude variation is known as the *beat frequency*.

The broken line in Fig. 6.5c has a frequency $(f_1 - f_2)/2$. But the maximum amplitude occurs twice in each of its cycles. So the frequency of the beats is given by

$$f_b = f_1 - f_2. \tag{6.4}$$

This is an important result.

(b) Measurement of frequency. The phenomenon of beats provides a very precise method of measuring the frequency f of a source, particularly for electromagnetic waves. For such waves we can produce a standard source whose frequency f_0 is known very precisely indeed. If we mix its output with that of the unknown and measure the beat frequency f_b, we have

$$f = f_0 \pm f_b. \tag{6.5}$$

The precision of the method lies in the fact that we can find a standard oscillator such that f_0 is close to f. Then f_b is a small quantity: so even a relatively imprecise measurement of it suffices to determine f precisely. This is an example of the Case I situation mentioned on p. 53.

Suppose, for example,

$$f_0 = 1\,000\,000 \text{ Hz} \quad \text{and} \quad f_b = 500 \pm 5 \text{ Hz}.$$

Then

$$f = 1\,000\,500 \pm 5 \text{ Hz}. \tag{6.6}$$

In other words, a measurement of f_b to 1 part in 100 has given the value of f to 1 part in 200 000.

To measure f_b it is necessary first to extract from y in Fig. 6.5c a signal equivalent to the positive envelope of the wave, a process known as *detection* or *demodulation* (Millman 1979). The frequency of the signal is then measured by some standard method, e.g. by converting the signal into a series of pulses, one for each cycle, and counting the number of pulses in a known time interval.

We have taken the positive sign - arbitrarily - in (6.5) to obtain the value of f in (6.6). How would we know which sign to take in practice? One way would be to measure f approximately to find out whether it was greater or less than f_0. In the present example a measurement precise to slightly better than 1 part in 2000 would be adequate for the purpose. Another way of selecting the correct sign is to make a slight change in f_0, say to increase it slightly, and observe whether f_b increases or decreases.

(*c*) *Sources of standard frequency*. It may be noted that we have not included an error for the value of f_0. This is because the frequency of a standard source may be known very precisely. Indeed, frequency - or its reciprocal, time - can be maintained and measured to a higher degree of precision than any other quantity in physics. (For a detailed account of frequency and time measurement see Kartaschoff 1978.)

Several methods are available for producing oscillations of a standard frequency. The simplest is to make use of a quartz crystal, which, by virtue of the piezoelectric effect, can act as a highly tuned circuit and produce an electromagnetic oscillation whose frequency is a characteristic of the geometry of the crystal and its elastic properties. Crystals giving a frequency stability of 1 part in 10^6 are readily available. If the temperature of the crystal is controlled, a stability of a few parts in 10^9 may be obtained.

Still higher stability is obtained from an oscillator whose frequency is controlled by transitions between two energy levels in an atom - usually rubidium or caesium. A rubidium oscillator is stable to about 1 part in

10^{11} and has the advantage that it can be built as a small portable unit. An oscillator controlled by a caesium beam in a resonant cavity gives the highest stability of all - about 1 part in 10^{13} - and the second is now *defined* in terms of one of the transitions of the caesium atom (see p. 195).

A number of radio stations throughout the world broadcast signals with the frequency of the carrier maintained by an atomic standard. In the United Kingdom, for example, there is a transmission from Rugby at a carrier-wave frequency of 60 kHz, stabilized to 2 parts in 10^{12}.* The signal may be fed into a laboratory oscillator of moderate stability, which is then stabilized to the same degree.

6.7 Negative feedback amplifier

(*a*) *Principle of negative feedback*. Suppose we have an electronic amplifier which has the property that, when a voltage V_i is applied across its input terminals, a voltage V_o appears across its output terminals, where

$$V_o = \alpha V_i. \tag{6.7}$$

We assume α is a constant and call it the *intrinsic gain* of the amplifier. We shall not concern ourselves with the internal details of the amplifier but shall simply represent it as a box with input and output terminals (Fig. 6.6).

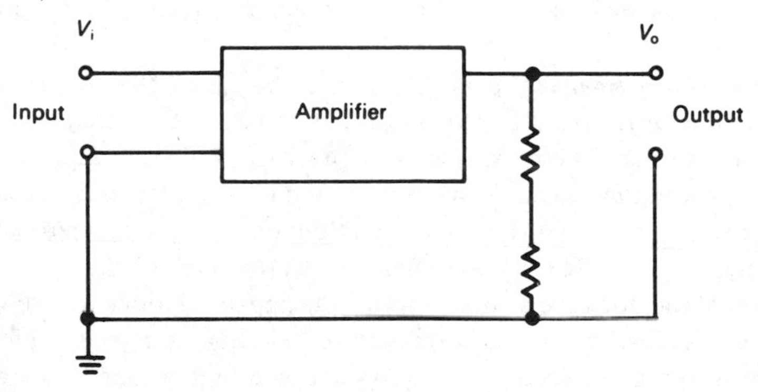

Fig. 6.6. Schematic representation of amplifier.

Suppose now that we have a certain signal voltage V_s that we wish to amplify, but that, instead of applying V_s directly to the input terminals, we subtract from it a fraction βV_o of the output and apply the remainder to the input terminals. This procedure of reducing the input signal by

* The short-term stability of the received signal may not be as good as that of the transmitted signal, owing to variations in ionospheric conditions.

something that depends on the output is known as *negative feedback.* Figure 6.7 is a schematic representation of the situation where the feedback signal is obtained from a simple resistance chain.

We have

$$V_i = V_s - \beta V_o. \tag{6.8}$$

Therefore

$$V_o = \alpha V_i = \alpha(V_s - \beta V_o), \tag{6.9}$$

whence

$$\frac{V_o}{V_s} = \frac{\alpha}{1 + \alpha\beta}. \tag{6.10}$$

So the net or overall gain is reduced by the feedback, since α and β are both positive numbers.

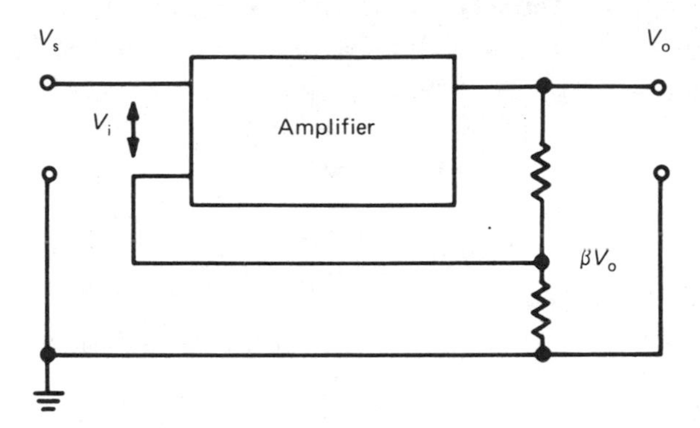

Fig. 6.7. Amplifier with negative feedback.

Now what is the point of this? The answer comes from considering what happens when $\alpha\beta$ is very much larger than unity. In that case we may neglect the 1 in the denominator in (6.10), and the gain becomes $1/\beta$. In other words, the net gain does not depend on the intrinsic gain of the amplifier, but only on the fraction β of the output that is fed back.

(b) Advantages of the method. (i) *Insensitivity to variations in supply voltage and amplifier components.* The quantity β can be fixed very precisely. It simply depends on having a pair of well-defined resistors. The quantity α on the other hand can vary for many reasons. For example, it may fluctuate owing to changes in the supply voltage. Again, it may vary from one amplifier to another owing to small changes in the various components – resistors, capacitors and transistors. Nevertheless, provided the value of β is the same for all the amplifiers, their overall gains will be almost equal. The fact that negative feedback reduces the gain of an amplifier is not really a disadvantage, because pure gain as such is quite easy to achieve.

Quite apart from (6.10), it is easy to see *qualitatively* why negative feedback reduces the sensitivity of the net gain to the intrinsic gain of the amplifier. Suppose the amplifier is working with feedback at a certain value of α, β, and V_s. Then V_o and V_i are given by (6.10) and (6.8). Now suppose that for some reason the value of α is reduced. This tends to reduce V_o, which reduces the feedback voltage βV_o, which results in an *increase* in V_i. Therefore the output V_o is not reduced as much as it would otherwise be. Of course it must be reduced by something, however small; otherwise the whole sequence of steps that we have just outlined would not occur at all.

Let us put in some numbers to show how negative feedback reduces the effect of a change in α on the net gain of an amplifier. If

$$\alpha = 20\,000 \quad \text{and} \quad \beta = \frac{1}{100}, \tag{6.11}$$

then

$$\text{net gain} = \frac{20\,000}{1+200} = 99{\cdot}50. \tag{6.12}$$

Now suppose that α drops to 10 000 – a very large drop indeed. The net gain becomes

$$\frac{10\,000}{1+100} = 99{\cdot}01. \tag{6.13}$$

So through the intrinsic gain changes by a factor of 2, the net gain changes by only $\frac{1}{2}$%.

(ii) *Improved frequency response.* Suppose the input is a sinusoidal voltage. For most amplifiers the intrinsic gain depends on frequency, owing to the various capacitances in the circuit. The advantage of obtaining the feedback voltage from a resistance chain is that β does not depend on frequency. So the overall gain of the amplifier is practically independent of frequency.

In a hi-fi audio amplifier, heavy negative feedback is used to achieve this situation, which means that musical notes of all frequencies, and their harmonics, are amplified by the same amount. This is necessary for the final result to be a faithful copy of the original, and the specification of the amplifier usually contains a statement about the constancy of the overall gain over a specified frequency range.

(iii) *Improved linearity*. If, for a signal of given frequency, V_o is not a linear function of V_i, we still write

$$V_o = \alpha V_i, \tag{6.14}$$

but α now varies with V_i. However, provided $\alpha\beta \gg 1$ for all values of V_i, the same reasoning applies as before, and

$$V_o \approx \frac{V_s}{\beta}. \tag{6.15}$$

So, provided β is constant and $\alpha\beta \gg 1$, V_o is approximately a linear function of V_s, and this is true no matter how non-linear the relation between V_o and V_i.

Other advantages of negative feedback are that the input impedance of the amplifier is increased and its output impedance decreased.

(*c*) *Stability*. If the sign of the feedback signal is reversed, so that it *adds* to the signal V_s instead of subtracting from it, we have *positive* feedback. Equation (6.10) still applies with β negative. It is now possible for $\alpha\beta$ to be equal to -1, in which case the equation gives an infinite value for V_o. What happens in practice is that V_o rises to a large value, the capacitive and inductive elements in the circuit give time delays, and the system becomes unstable and starts to oscillate. Circuits based on this principle are in fact designed deliberately as oscillators.

However, if the circuit is required as an amplifier, such behaviour is highly undesirable. The danger with any type of feedback circuit is that even though we intend always to have negative feedback, we may in fact get positive feedback at certain frequencies. This is because the amplifier produces phase shifts in the signals and these shifts depend on frequency. Part of the art of designing an amplifier is to ensure that it remains stable at all frequencies. The theoretical conditions for this have been worked out by Nyquist and others.

For a full discussion of the theory and design of negative feedback amplifiers you are referred to one of the books listed on p. 205. In this section we have given only an introduction to this important idea, which may be applied to all amplifying devices.

6.8 Servo systems

(*a*) *The servo principle*. Suppose we have an apparatus S, some feature of which we wish to control from a unit C. We may take a control signal from C, put it through an amplifier A and apply the output to S.

Suppose now we allow S to produce a signal F that is a measure of the quantity we are trying to control, feed it back so as to subtract from the control signal and apply the difference to the input terminals of the amplifier – Fig. 6.8. This system of controlling the apparatus is known as a *servo* system. We see that it is based on the idea of negative feedback, and many of the advantages of negative feedback that we considered in the last section in relation to amplifiers apply also to quite general servo systems.

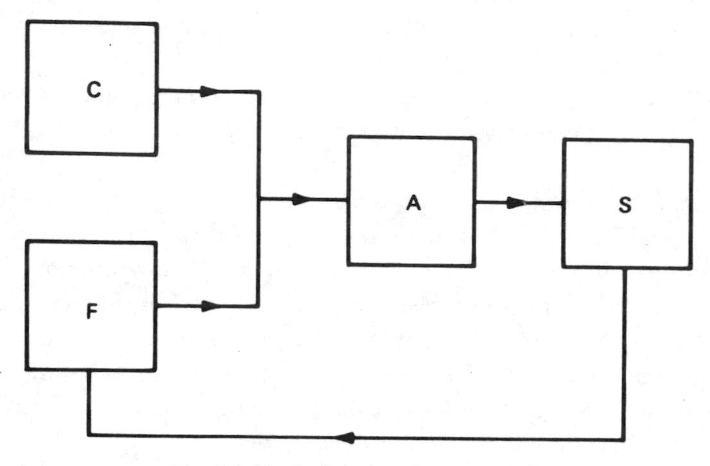

Fig. 6.8. Block diagram of servo system.

(b) Example – temperature control. As an example of the servo principle we consider a bath whose temperature T we require to be kept constant, at a value that depends on the setting of a control. The bath B – Fig. 6.9 – loses heat to its surroundings, and its temperature is maintained by means of a heating coil H. The current through H is supplied by a variable autotransformer V whose output, rather like that of a potentiometer, depends on the position of a movable contact. The magnitude of the heating current thus depends on the position of the contact, and this is changed by a dc electric motor M via a reduction gear R. The speed of the motor depends on the current through its field coils, is zero when this current is zero and reverses when the current reverses.

Let the control C produce a voltage V_c. The feedback is provided by some device that gives a signal that depends on the temperature of the bath, for example a thermocouple D. Suppose this produces a voltage V_f. The voltage $V_c - V_f$ is fed into the amplifier A, the output of which is applied to the field coils of the motor. The connections are made so

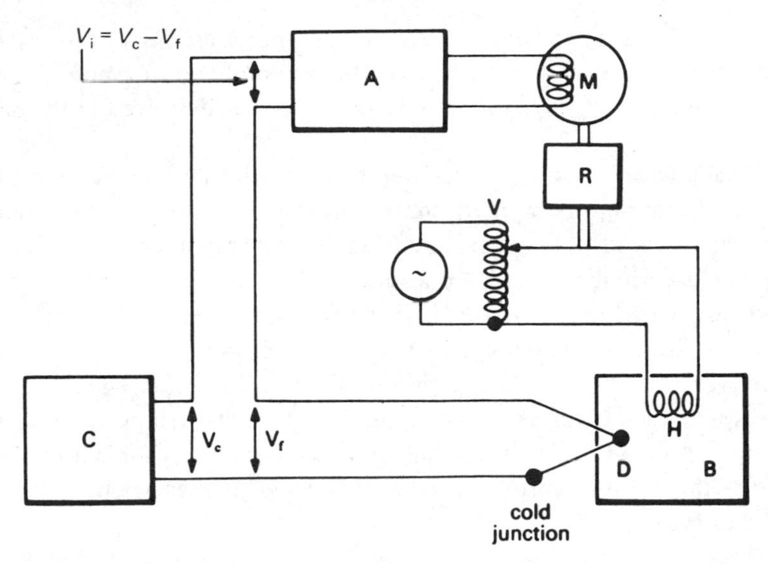

Fig. 6.9. Servo system for maintaining bath at constant temperature.

that when $V_c - V_f$ is positive, the motor drives the autotransformer in the direction that increases the heating current in H.

The required temperature T is that for which the thermocouple output V_f is equal to V_c. When the bath is at this temperature, the signal supplied to the amplifier is zero; the motor is therefore stationary, and the autotransformer setting remains fixed. The heat supplied to the bath under these conditions is just equal to the heat losses at temperature T. If for any reason the temperature should fall, for example because the heat loss increases due to a fall in the surrounding temperature, V_f falls, and the positive signal $V_c - V_f$ acts via the amplifier, motor, and autotransformer to increase the power supplied. Conversely, if the temperature rises above T, $V_c - V_f$ becomes negative, and this results in a decrease in the power supplied.

We see then the essential features of a servo system. We must have a control signal V_c. This acts as a reference. We must have an output signal V_f, which is a measure of the physical quantity that is being controlled. The difference between the two signals is used to operate a device which acts so as to reduce this difference to zero. In short a servo system provides *automatic control.*

The constant temperature bath illustrates some of the features of negative feedback mentioned in the previous section. For example, the functional relation between V_c and T is almost the same as that between

V_f and T. It depends hardly at all on the other relations in the system, such as that between the input signal to the amplifier and motor speed, or between the autotransformer setting and power delivered to the bath.

(c) *Stability*. So far we have ignored time lags in the system. However, these have an important effect on its behaviour. Suppose in the example we move the control to correspond to a new higher temperature T'. While the bath is warming up and T is less than T', the autotransformer control is being pushed up to provide higher and higher heating currents. This continues until the thermocouple reaches T'. However, owing to thermal time lags in the system, heat continues to come in at a rate greater than the equilibrium heat loss at temperature T'. So the temperature of the bath rises above this value. Whereupon the servo mechanism starts to reduce the heating current, and, again owing to thermal time lags, it reduces it too much.

This type of oscillatory behaviour in a servo system is known as 'hunting'. It is analogous to the instability we mentioned previously in connection with positive feedback in amplifiers. Mathematically the two types of behaviour are equivalent. Instability in servo systems is overcome by damping, that is, by diminishing the change in the applied quantity - heat into the bath in the present example - in various ways. Just as in feedback amplifiers, the avoidance of instability is one of the main problems in the design of servo systems.

6.9 Natural limits of measurement

It might be thought that, if we used sufficiently sensitive instruments and took enough care, we could make measurements as precisely as we pleased. But this is not so. Quite apart from the limitations imposed by the uncertainty principle in quantum physics, with which we are not here concerned, there are several phenomena which give rise to random fluctuations in measuring devices. These fluctuations are known as *noise* and provide natural limits to the precision that can be achieved.

(a) *Brownian motion*. One source of random fluctuation in a measuring instrument is Brownian motion; an example is provided by a small mirror suspended by a torsion fibre. If the restoring couple due to an angular displacement θ is $c\theta$, then the potential energy is

$$V = \tfrac{1}{2}c\theta^2. \tag{6.16}$$

The mirror is being constantly bombarded by gas molecules, and, though the value of the couple they exert is zero when averaged over time, its instantaneous value is not, and the mirror undergoes random fluctuations about its mean position.

The mirror is in thermal equilibrium with the gas molecules. So, by the theorem of the equipartition of energy, the average value of V is $\frac{1}{2}kT$, where k is the Boltzmann constant and T the absolute temperature.* The average value of θ^2 is therefore

$$\overline{\theta^2} = \frac{kT}{c}. \tag{6.17}$$

Since $k = 1{\cdot}38 \times 10^{-23}\,\mathrm{J\,K^{-1}}$, the effect is usually small at room temperature. However, for small values of c, the fluctuations become appreciable; in fact the phenomenon has been used to provide a value for k (Kappler 1938).

Notice that the effect cannot be reduced by reducing the pressure. Such a reduction decreases the number of molecules striking the mirror per second and changes the motion from a high speed jitter to one with sinusoidal swings, but the value of $\overline{\theta^2}$ remains the same. Some diagrams of the motion are given in Fowler 1936, p. 783.

(*b*) *Johnson noise.* A second and important type of thermal agitation is the phenomenon known as *Johnson* or *thermal noise.* A resistance R at temperature T is found to act as a generator of random emfs. They may be regarded as arising from the thermal motion of the conduction electrons. The resulting emf E varies in an irregular manner, and, if a frequency analysis is made, the average value of E^2 in the frequency range f to $f + \mathrm{d}f$ is

$$\overline{E^2} = 4RkT\,\mathrm{d}f. \tag{6.18}$$

We are simply quoting this result.† It follows from basic thermodynamic reasoning and is valid provided $f \ll kT/h$, where h is the Planck constant. (At room temperature $kT/h = 6$ THz.)

Notice that the right-hand side of (6.18) does not contain f itself. In other words, except at very high frequencies, Johnson noise is constant throughout the frequency spectrum and is proportional to the product RT. For this reason, when the signal to be amplified is very weak and the noise therefore very troublesome, the amplifier is sometimes cooled

* The equipartition theorem is discussed in most textbooks on statistical mechanics; see for example Reif 1965.

† For comprehensive accounts of noise in measurements see Robinson 1974 and van der Ziel 1976. Both books give a proof of (6.18). For a good introduction to frequency (Fourier) analysis see Gough, Richards and Williams 1983.

to liquid-air and even liquid-helium temperatures. This not only reduces the value of T but also that of R.

(c) *Noise due to the discreteness of matter.* Electric current is carried by discrete particles - electrons and electron holes. In successive intervals of time the number of particles in motion fluctuates, a phenomenon known as *shot noise.* The smaller the current, the larger the fractional fluctuation. Examples of shot noise are the fluctuations in the current of a semiconductor diode due to variations in the rate at which the charge carriers cross the p–n junction, and fluctuations in electron emission by a thermionic cathode or photocathode.

(d) *Flicker noise.* In addition to Johnson noise and shot noise, which are fundamental and do not depend on the quality of the components in which they occur, there is another type of noise which depends on the detailed properties of the component. It is known as *flicker* or *$1/f$ noise* because its frequency spectrum varies as $1/f$. Thus it is most important at low frequencies. The physical origin of flicker noise varies from one device to another. It commonly arises from variations of properties with time in devices which are not in thermal equilibrium, but are subject to an external disturbance such as a bias or signal voltage. For example, a carbon resistor consists of a large number of small granules with contact resistance between them. When a current passes, small random motions of the granules produce changes in the overall resistance, and the voltage across the resistor fluctuates with time. In a biased semiconductor device, flicker noise arises from fluctuations in the rates at which the majority and minority carriers are generated and recombine in heterogeneous regions of the crystal.

(e) *Noise in general.* Except for work involving the detection of very weak signals, the various sources of noise we have been describing are not usually limiting factors in normal laboratory measurements. Other sorts of disturbance are often present and may be more serious. Common examples are pickup from the mains, interference by nearby electrical machinery, spurious signals due to bad electrical contacts and faulty electronic components.

Exercises

6.1 A stroboscope is a device for measuring the frequency f of a rotating object by viewing it with a flashing light of known frequency f_0 and measuring the apparent frequency f_{app} of rotation.

(a) Show that if f is roughly equal to mf_0, where m is an integer,

$$f_{\text{app}} = f - mf_0.$$

What is the significance of a negative value for f_{app}?

(b) If

$$m = 5,$$
$$f_0 = 100{\cdot}00 \pm 0{\cdot}01 \text{ Hz},$$
$$f_{\text{app}} = 0{\cdot}40 \pm 0{\cdot}05 \text{ Hz},$$

calculate the value of f and its standard error.

6.2 The emf E of a Weston standard cell at 20 °C is 1·0187 V. The temperature variation of E at this temperature is

$$\frac{\mathrm{d}E}{\mathrm{d}T} = -37\ \mu\text{V K}^{-1}.$$

It is required to measure ΔE, the difference in E at 20 °C and 30 °C for a particular cell S_1. Consider the following two methods:

(a) E is measured with a precision potentiometer at 20 °C and 30 °C. The precision of each measurement is 10 μV.

(b) The difference between the emfs of S_1 and another similar standard cell S_0 is measured by the method shown in Fig. 6.10. The standard cells are connected with their emfs opposed. R is a standard resistance of 0·001 Ω, known to 1 part in 10^3. The resistance P is varied until the deflection of the galvanometer G is zero. The current through the milliammeter A is then read with a precision of 1%. The measurements are made with S_1 first at 20 °C and then at 30 °C, S_0 being kept at 20 °C throughout.

Compare the precisions of the two methods of measuring ΔE. (Assume that the smallest voltage the galvanometer can detect is 0·2 μV.)

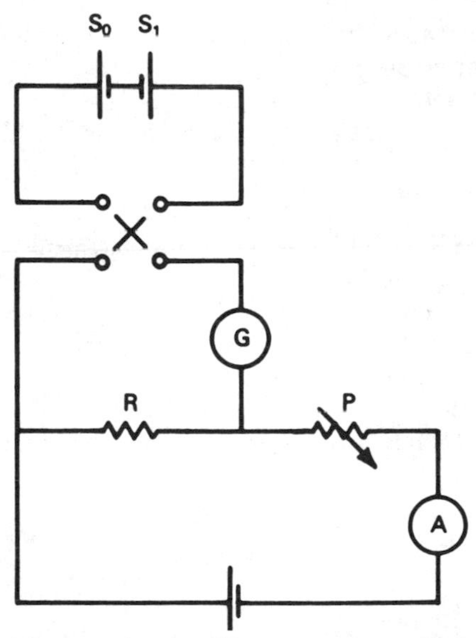

Fig. 6.10. Circuit for measuring the difference in the emfs of two standard cells.

6.3 Make a critical comparison of the following types of thermometers on the basis of range of temperature, precision, convenience (including situations of particular applicability), and cost:
(a) mercury in glass,
(b) thermocouple,
(c) platinum resistance,
(d) thermistor,
(e) constant volume gas thermometer,
(f) optical pyrometer.

6.4 Make a list of possible ways of measuring magnetic fields and compare them as in the previous exercise.

The following discussion exercises are meant to make you think about the nature of measurement and to help you see how measurement and theory are inextricably bound together. Some of them involve ideas beyond the scope of the average first-year undergraduate course.

6.5 Discuss the concept of size as applied to
(a) an atom,
(b) an atomic nucleus.

6.6 Discuss appropriate methods of measuring the distance between
(a) atoms in a crystal,
(b) the two atoms in a hydrogen molecule,
(c) two points about 10 km apart on the Earth's surface,
(d) the Earth and the Moon,
(e) the Earth and a nearby star,
(f) the Earth and a distant star.

6.7 Discuss appropriate methods of measuring the mass of
(a) a sack of potatoes,
(b) a bar of gold,
(c) a proton,
(d) a neutron,
(e) the Earth.

6.8 Explain what is meant by the following statements and how they may be verified:
(a) The temperature of a salt following adiabatic demagnetization is 0·001 K,
(b) The temperature in a plasma is 50 000 K,
(c) The temperature in outer space is 3 K,
(d) The temperature of a certain nuclear spin system is negative.

7

An analysis of some experiments

The aim of the present chapter is similar to that of the last, namely to consider precise and accurate methods of measurement. The main difference is that in this chapter we shall look at the methods in the context of complete experiments.

Four experiments have been chosen. They are all well described in standard textbooks or original papers, to which you are referred for more details. We analyse them here with the emphasis on the way the errors have been reduced or eliminated.

7.1 Comparison of low resistances by means of a potentiometer

(a) Description of the method. A simple potentiometer circuit is shown in Fig. 7.1. The (rechargeable) battery E sets up a potential difference across the slide wire AC. The two resistors, whose resistances R_1 and R_2 are to be compared, are connected in series with the battery F. Since the same current flows through them, the potential differences across them, V_1 and V_2, satisfy the equation

$$\frac{V_1}{V_2}=\frac{R_1}{R_2}. \tag{7.1}$$

The points Y and Z are connected respectively to a and b, and the slider B is moved until no current passes through the galvanometer G. The value of AB, x_1, is read from the potentiometer scale. The measurement is repeated with Y and Z connected to c and d, and the second value of AB, x_2, is found. Each value of x is proportional to the potential difference across the corresponding resistor. Therefore

$$\frac{R_1}{R_2}=\frac{V_1}{V_2}=\frac{x_1}{x_2}. \tag{7.2}$$

The purpose of the variable resistors P and Q is to ensure that the balance points for both R_1 and R_2 are on the scale AC, with the larger one fairly close to C.

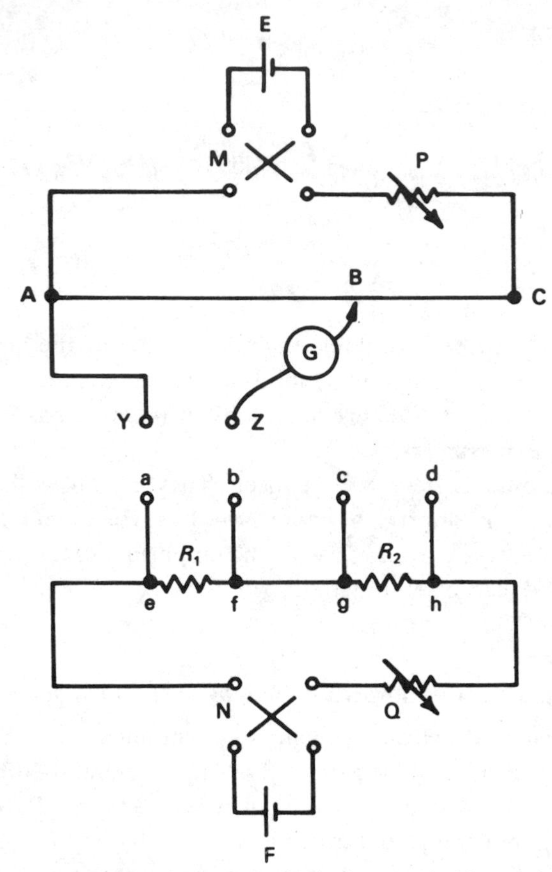

Fig. 7.1. Circuit for potentiometer method of comparing low resistances.

***(b)* Sources of error.** We list the major sources of error in obtaining R_1/R_2 from the simple procedure described above.

(i) The potentiometer scale may not be correctly marked.

(ii) The current through the wire AB may not be the same for the two measurements; in which case the relation

$$\frac{V_1}{V_2}=\frac{x_1}{x_2}$$

is not true.

(iii) The current through R_1 and R_2 may not be the same when the two measurements are made; in which case the relation

$$\frac{R_1}{R_2}=\frac{V_1}{V_2}$$

is not true.

(iv) The zero on the potentiometer scale may not correspond to zero potential.

(v) The slide wire AC may not be uniform; in which case again the relation

$$\frac{V_1}{V_2} = \frac{x_1}{x_2}$$

is not true.

(vi) The resistances R_1 and R_2 may not be reproducible, but may vary slightly according to the way the connections are made at e, f, g and h.

(vii) The temperatures of the resistors may not be the same.

(viii) There may be sources of emf other than E and F in the two circuits.

(c) *Elimination of errors.* We consider these effects in turn and see how the errors to which they give rise may be reduced or eliminated.

(i) For an instrument such as this it is worth using a precision steel rule for the potentiometer scale. Its error is probably negligible and can certainly be made so by comparing the rule with a standard scale.

(ii) and (iii) The currents through AB and R_1 and R_2 depend on the emfs of the batteries E and F. These should be fairly constant provided

(a) the batteries are reasonably well charged,

(b) the two circuits remain closed all the time, so that the two batteries have settled down to a steady rate of discharge,

(c) the measurements of x_1 and x_2 are made in fairly quick succession.

The fact that the emfs are not truly constant with time means that we do not have an exactly steady state. Whenever two quantities are being measured under these circumstances, it is a good principle to make what may be called 'sandwich' measurements. In this case the procedure would be to measure x_1, then x_2 and then x_1 again. The average value of x_1 is taken. We may extend the idea and continue to alternate the measurements of x_1 and x_2. Provided we start and finish with the same quantity, i.e. either x_1 or x_2, and in addition make the measurements frequently and at roughly constant intervals, the error due to *steady* changes in the emfs of E and F becomes very small indeed.

(iv) It is not easy to make the zero of the scale correspond exactly to zero potential difference across AB. There is likely to be a zero error, a constant quantity δ that must be added to each slider reading x in order to obtain the correct value. The best method of obtaining δ is to disconnect the battery F, thereby giving the potentiometer zero potential difference to measure. The balance point is just beyond the extreme limit of the slider. So we plot the actual galvanometer deflection against x for a few settings of the slider close to its limit and extrapolate to zero deflection. This value of x is equal to $-\delta$ (Fig. 7.2).

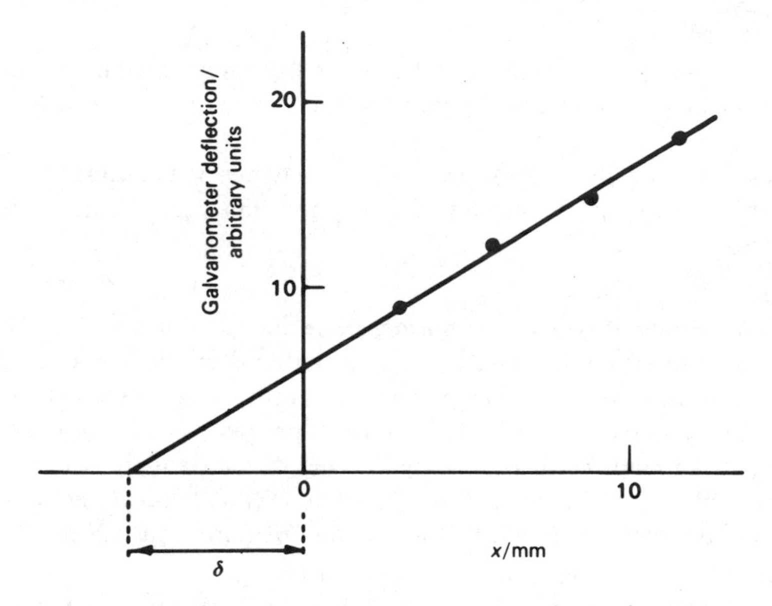

Fig. 7.2. Measurements to determine zero error of potentiometer.

(v) Provided the wire has not been ill-treated, the error due to non-uniformity is probably fairly small. If we wish to test this it would *not* be good practice to measure the diameter of the wire as a function of length. In general one should always try to measure the quantity that is specifically required, in this case the potential difference as a function of length. So the best way of testing the uniformity of the wire is to use a pair of sliders set a constant distance apart, and to measure the potential difference across them as they are slid along the wire AC.

(vi) The next point to consider is the reproducibility of the resistances R_1 and R_2. One of the advantages of the potentiometer method of comparing resistances is that the resistance of the potential leads, i.e. ae, bf, etc., in Fig. 7.1, is irrelevant because they carry no current at the balance point, and hence there is no potential drop along them. However, if the resistances R_1 and R_2 are small - 10^{-3} to $10^{-4}\,\Omega$ - it is important to consider how the potential and current leads are connected. This is because a small resistance – known as a *contact* resistance, and of the order of $10^{-6}\ \Omega$ – is always introduced at a terminal. These resistances vary in magnitude, and it is essential that they should not be included in the quantity actually measured.

Suppose that the resistor only has two terminals and we attach a current and a potential lead to each terminal - see Fig. 7.3a, which shows a screw terminal with the spade ends of the two leads. This is really equivalent to the situation shown in Fig. 7.3b, where r_a and r_b are the contact resistances. If the current lead is the upper one at each terminal, the equivalent circuit is as shown in Fig. 7.3c. The contact resistances r_b and r_d are included with R_1 in the measurement. If the potential terminals are uppermost, it is still true that the contact resistances in the positions of r_b and r_d are included.

The difficulty is overcome by having *four* terminals on the resistor as in Fig. 7.4. The current leads are attached to A and B, and the potential leads to C and D. The contact resistances now have no harmful effect. Those at A and B only diminish the current by a trivial amount, but we are only interested in its constancy, not its actual value. Those at C and

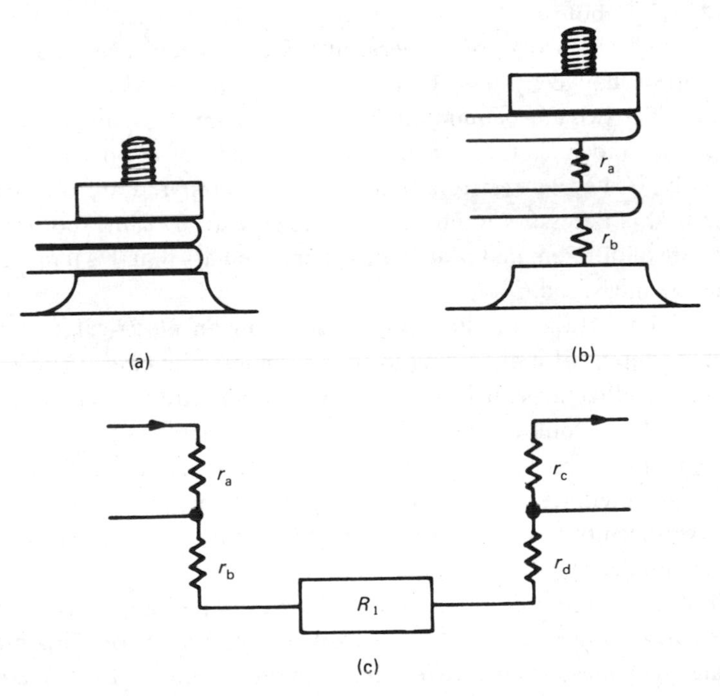

Fig. 7.3. Resistor with two terminals – the effect of contact resistances.

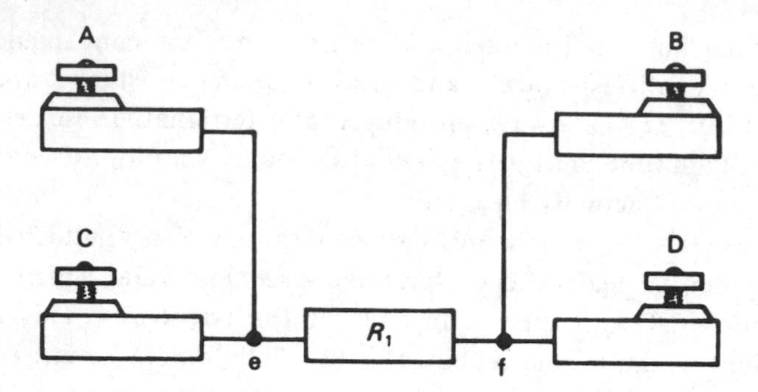

Fig. 7.4. Resistor with four terminals.

D give no error, because no current passes through them at the balance point. The device therefore actually gives the resistance between the points e and f. The wires are soldered at these points once and for all, so the resistance measured is a fixed quantity. Notice that if we use only two of the four terminals for the current and potential leads, the entire advantage is lost, and we are back to the situation of Fig. 7.3c.

(vii) The resistivity of most substances varies with temperature. At room temperature the resistivity of many metals, including copper, increases by about 0·4% for a temperature rise of 1 K. This is a large effect for precise work, but several alloys have been developed whose resistivities change by less than 1 part in 10^5 per kelvin at room temperature. The two most commonly used are manganin, an alloy of copper, manganese and nickel, and constantan, an alloy of copper and nickel.

Clearly, if the two resistors that are being compared are of the same material, it is necessary to ensure that they are at the same temperature. If they are of different material, it is further necessary that this temperature should be measured.

(vii) All junctions of different materials in an electrical circuit are possible sources of emfs owing to the thermoelectric effect. Those most likely to produce an emf in practice are copper wires attached to brass terminals. If we look at the basic circuit in Fig. 7.1 we can see that any extra emf in series with E or F does not affect the result, because, as previously noted, we are not interested in the actual values of the currents in the two circuits, but only in their constancy while the two measurements of x are made.

However, in the measurement of x_1 for example, any source of emf in the galvanometer circuit AYaefbZBA will give an error. This may be eliminated by measuring x_1 twice, first with the currents in the two circuits in one direction and then with them both reversed by means of the reversing switches M and N. The thermoelectric emfs remain in the same direction. The mean of the two values of x_1 is the value that would be obtained if the extra emfs in the galvanometer circuit were zero. The same pair of measurements is made for x_2.

(d) Comments on the method. (i) *Basic advantage.* The advantage of the method is that the resistance measured is that of R_1 and R_2 alone, without any correction for the leads. So it is particularly useful for low resistances (of the order of 10^{-1} to $10^{-4}\,\Omega$), and for situations where the resistor is inaccessible, which is often the case in experiments where it is at a very high or very low temperature. The main disadvantage is the need to keep the currents in the two circuits fairly constant. This necessitates rapid and/or repeated measurements.

(ii) *Precision.* The precision of the method as described here is limited by the slide wire, the length of which is usually 500 mm. The balance point can be found and the reading made with a precision of about 0·1 mm. So the overall precision is, at best, about 2 parts in 10^4. To improve on this it is necessary to replace the slide wire by a set of coils as in a commercial potentiometer. A precision of a few parts in 10^5 is then possible.

In order to achieve this very high precision further precautions must be taken. To start with, the resistor must be made of a substance like manganin, and it must be prepared and maintained under standardized conditions of strain, temperature, and humidity. Second, the variation with time of the emfs of the cells E and F must be considerably reduced by means of some form of stabilizing device. A simple example is shown in Fig. 7.5. The cell G has an emf of 12 V - compared with the 2 V of

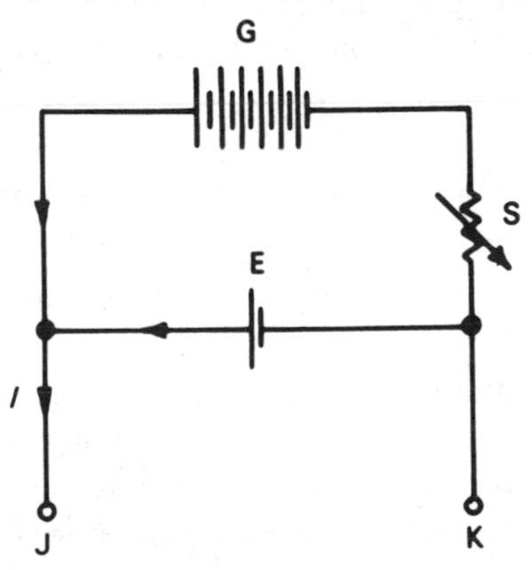

Fig. 7.5. Simple method of stabilizing the emf of the cell E.

E. The resistance S is adjusted so that, of the current I passing into the potentiometer, only say 5% comes from E and the remaining 95% from G. The voltage across the points JK is determined almost entirely by that of the battery E, but since the latter is passing very little current its voltage changes extremely slowly.

(*e*) *Null methods*. One of the reasons for discussing the potentiometer is that it is a good example of a null method. A *null* method is one in which the quantity X being measured is opposed by a similar quantity Y, whose magnitude is adjusted until some indicating device shows that a balance has been achieved. This is to be contrasted with a *direct* method in which the quantity being measured produces a deflection or reading on an instrument.

Null methods have several important advantages over direct methods. The adjusted quantity Y in a null method tends to be more stable and reproducible than the instrument in a direct method, and its value can usually be read more precisely. The instrument in a direct method must be calibrated and should preferably be a linear device. By contrast, the indicating device in a null method only has to show when some quantity is zero. Therefore it does not have to be calibrated and need not be linear, though it is convenient if it is so in the neighbourhood of zero. The main property of interest in the device is its sensitivity, that is, the minimum out-of-balance effect that it can detect.

A disadvantage of null methods is that they tend to be slower than direct ones, since the balance point is approached by stages. This can be overcome at the expense of further complication by using a servo system to find the balance point automatically. It is important to realize that null methods can give only relative, and not absolute, values. So an essential feature is the existence of a precisely known standard.

7.2 Alternating current method of measuring resistivity

(*a*) *Introduction*. Although the potentiometer is an instructive device, it suffers from the disadvantage that it is a dc instrument in which thermoelectric and drift effects give errors that are troublesome to eliminate. For this reason ac methods are now more commonly used for precise electrical measurements. As an example of a modern technique we have selected a method developed by Friend and others for measuring the resistivity and the Hall effect in metallic specimens at low temperatures. The design of the apparatus contains a number of interesting features showing what can be done with present-day semiconductor devices. We give here a simplified account of the method. Further details can be found in Friend and Bett 1980. To follow the discussion you will need some basic ideas on devices such as operational amplifiers, binary counters, and so on. These may be found in a number of books on electronics - see for example Horowitz and Hill 1980.

(b) Description of the method. The basic experimental arrangement is shown in Fig. 7.6. A sine-wave voltage is generated by digital means at a frequency of $f_0 = 70$ Hz and fed into a unit which produces an alternating current of the same frequency and of constant amplitude. The current I is passed through the sample via a pair of contacts A and B. The resulting voltage V developed across another pair of contacts C and D is fed into a unit, known as a *lock-in amplifier*, which acts as a highly tuned detector or demodulator, and gives a dc output proportional to the amplitude of the alternating voltage V. An essential feature of a lock-in amplifier is the injection of a reference signal with the same frequency f_0 as the signal from the sample. This is provided by the sine-wave voltage generator.

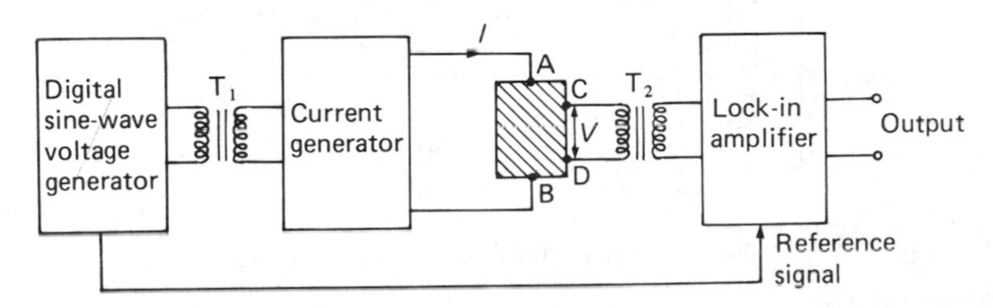

Fig. 7.6. Block diagram of apparatus for measuring the resistivity of a metallic sample (shown shaded).

The current I is determined in a preliminary experiment by passing it through a known resistance and measuring the voltage across the resistance. The lock-in amplifier is calibrated by applying a known alternating voltage to its input terminals and measuring the resulting dc voltage at the output. These measurements are made with a digital voltmeter which can measure dc and ac voltages and also resistance.

To obtain the resistivity of the sample, the measurements are repeated with A and D as the current contacts and B and C as the voltage contacts. An ingenious theorem by van der Pauw (1958) shows that, for a sample in the form of a lamina of uniform thickness, the two sets of measurements, plus the thickness of the lamina, suffice to determine the resistivity of the material – irrespective of the shape of the lamina and the location of the points A, B, C, D.

With this apparatus the experimenters were able to measure the current and voltage with a precision of about 1 part in 10^4. We consider each unit of the apparatus in turn to show how it functions.

(c) *Digital sine-wave voltage generator*. The elements of this unit are shown in Fig. 7.7. A timer circuit produces clock pulses at a frequency of $256f_0$. These are fed into an 8-bit binary counter which produces the numbers 0 to 255 in succession. The entire sequence is thus repeated with a frequency f_0. The numbers are fed into a read-only memory with 256

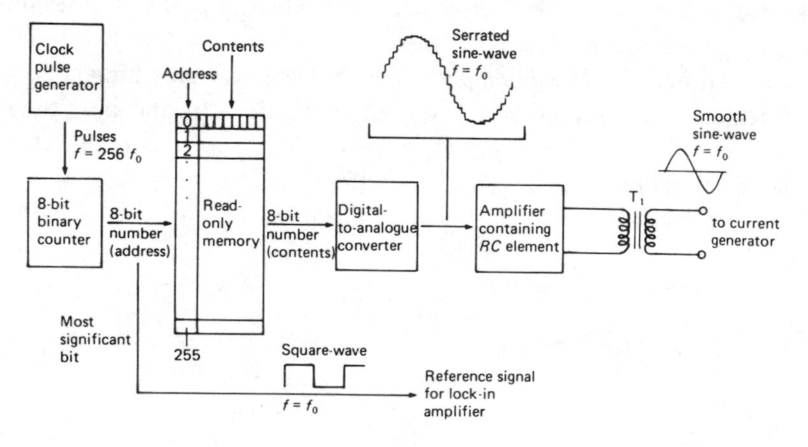

Fig. 7.7. Digital sine-wave voltage generator.

registers. Each number locates a register, the contents of which – also an 8-bit number – are produced at the output of the memory unit. The contents of register n are preprogrammed to be

$$c_n = \text{nearest integer to } \left\{\left[\sin\left(\frac{2\pi n}{256}\right)+1\right]\times 127{\cdot}5\right\}. \qquad (7.3)$$

The successive integers c_n (which you can see are in the range 0 to 255) are fed into a digital-to-analogue converter. This is a circuit that produces a voltage proportional to the digital input (Horowitz and Hill 1980, p. 410). The output from the convertor is thus a serrated sine-wave as shown schematically in Fig. 7.7. The serrations are removed by an *RC* element in the circuit. Finally a 1:1 transformer T_1 removes the dc component, thereby providing a floating voltage signal which acts as the input to the current generator. This signal is a smooth sine-wave of frequency f_0. Its amplitude may be varied in steps from 0·5 V to 2·5 V.

The reason for using a digital method to produce the sine-wave voltage is that the amplitude of the wave is very stable, being fixed by the digital-to-analogue converter. This unit has a built-in voltage reference – see section (g) – that regulates the output voltage to a few parts in 10^5.

The most significant bit of the 8-bit number from the binary counter is used as a reference signal for the lock-in amplifier. It is a square-wave with frequency f_0.

(d) Current generator. The current generator is shown in the left half of Fig. 7.8. The operational amplifier A_1 has the property that its gain is very high, so the voltage across its input terminals is very small and may be taken to be zero. The voltage across the resistor R_1 is thus equal to

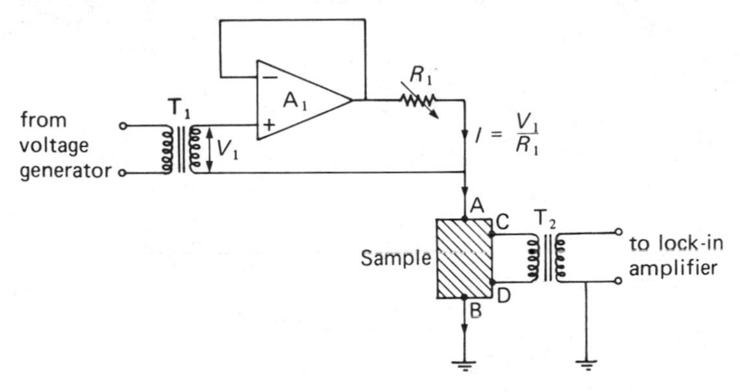

Fig. 7.8. Current generator and sample.

the voltage V_1 from the secondary of the transformer T_1. The current through R_1 is therefore

$$I = \frac{V_1}{R_1} \tag{7.4}$$

Since the input impedance of the operational amplifier is high and the voltage across the input terminals is small, the current through the input terminals is very small indeed. Therefore virtually all the current I passes through the sample via the contact points A and B. The resistance R_1 may be varied in steps to give a range of current values from 10 μA to 20 mA.

The basic property of the current generator is that the current depends only on the values of V_1 and R_1 according to (7.4). It does not depend on the resistance of the sample or on any contact resistances. So the values of the current, obtained in the calibration measurements when the current is passed through a standard resistor, remain valid when the resistor is replaced by the sample.

The voltage developed across the points C and D is applied to a step-up transformer T_2 with a 1 : 100 ratio, which serves two purposes. The first is to match the low impedance of the sample to the higher impedance required by the lock-in amplifier. The second is to ensure that the signal applied to the amplifier depends only on the voltage difference $V_C - V_D$, and not on the common-mode signal, i.e. the mean voltage $\frac{1}{2}(V_C + V_D)$.

(e) Lock-in amplifier. A simplified version of this unit is shown in Fig. 7.9. M_1 and M_2 are a pair of MOSFETs,* which in the present context act simply as a pair of switches. Each switch is closed when the gate of the MOSFET is at a positive voltage and open when the gate voltage is

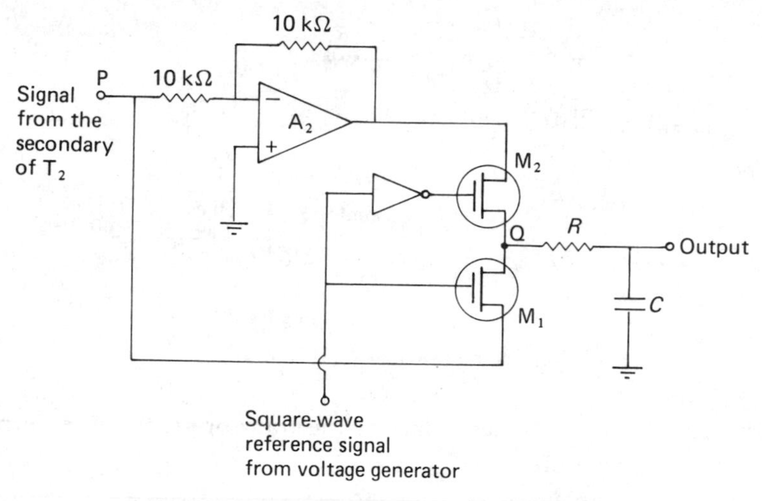

Fig. 7.9. Simplified version of lock-in amplifier.

negative. The square-wave reference signal from the voltage generator is applied directly to the gate of M_1 and, via an inverter, to the gate of M_2. Thus when the reference signal is positive, M_1 is closed and M_2 is open; when the reference signal is negative the reverse is true.

When M_1 is closed, the signal at P from the secondary of the transformer T_2 appears unchanged at Q. When M_2 is closed, the signal appears at Q with its sign reversed. This is because the non-inverting terminal of the operational amplifier A_2 is earthed. Since the voltage across the input

* A MOSFET is a metal-oxide-semiconductor-field-effect transistor. See Horowitz and Hill, 1980, p. 245 for a description of the device.

terminals is close to zero, the voltage of the inverting terminal is effectively zero – it is a 'virtual earth'. The potential differences across the two 10 kΩ resistors are equal. Therefore, since their common point is at earth potential, the voltages P and Q are equal and opposite. The signal at Q is thus the rectified form of the signal at P – Fig. 7.10. The *RC* circuit on the right-hand side of Fig. 7.9 acts as a low-pass filter i.e. the voltage across the capacitor C is, apart from a small ripple, equal to the dc component of the signal at Q, which in turn is proportional to the amplitude of the alternating voltage at P.

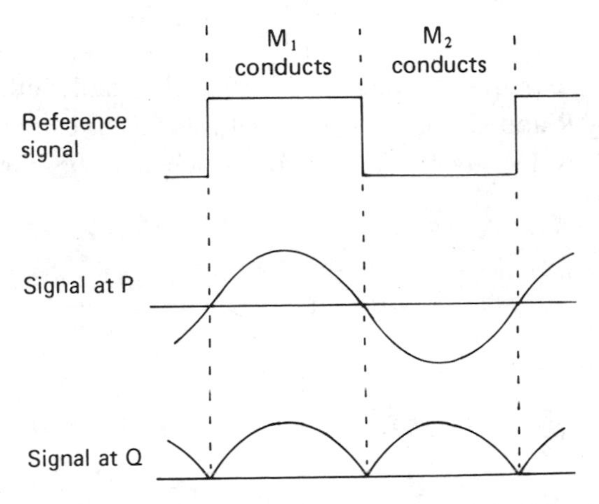

Fig. 7.10. Waveforms in lock-in amplifier. The points P and Q are indicated in Fig. 7.9.

(f) Noise reduction. In a typical set of measurements the resistance of a small metallic sample varies from about 10^{-3} Ω at room temperature to about 10^{-6} Ω at low (liquid helium) temperature. To avoid heating the sample, the maximum current that can be used is about 1 mA. The signal from the sample is therefore of the order of nV. The ac method eliminates the effects of thermoelectric voltages which, in the present experiment where leads are taken from room temperature to liquid-helium temperature, are of the order of μV. However, with such a small signal, noise is a major problem. Apart from Johnson and $1/f$ noise mentioned in the last chapter, there are effects due to mechanical vibrations of the equipment in the presence of magnetic fields. In general there is also pickup from the mains at a frequency of 50 Hz, but the experimenters avoided this by working at a frequency of 70 Hz.

The purpose of the lock-in amplifier and the low-pass filter, represented by the *RC* element, is to reduce the noise. Consider the *RC* element first - Fig. 7.11a. For a sinusoidal wave of frequency f the relation between the output V_C and the input V_Q is (see exercise 7.2)

$$\left|\frac{V_C}{V_Q}\right| = \frac{1}{(1+4\pi^2 f^2 C^2 R^2)^{1/2}}. \tag{7.5}$$

The relation is plotted in Fig. 7.11b. If we define a frequency f_b by the relation

$$2\pi f_b CR = 1, \tag{7.6}$$

we can see that

$$\text{for } f < f_b, \quad \left|\frac{V_C}{V_Q}\right| > \frac{1}{\sqrt{2}} = 0{\cdot}707. \tag{7.7}$$

We may take the frequency range $0 < f < f_b$ as the bandwidth of the filter. The values of R and C are chosen so that $f_b \ll f_0$. For example, a typical value of RC is 1 s, giving $f_b = 0{\cdot}2$ Hz, which may be compared with $f_0 = 70$ Hz.

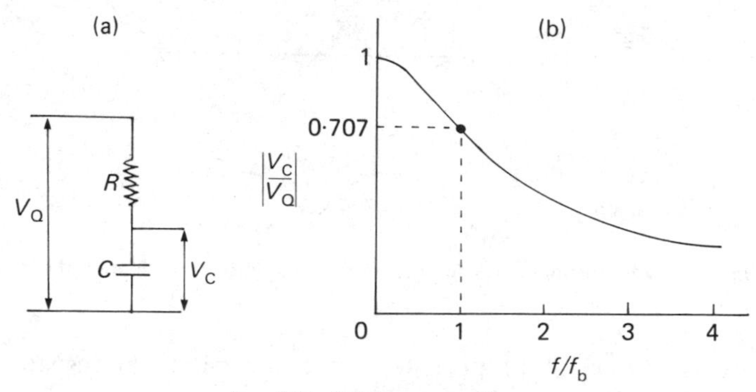

Fig. 7.11. *RC* low-pass filter.

Now consider the lock-in amplifier, which acts as a mixer for the reference and the signal voltages. The reference voltage has a square waveform of fundamental frequency f_0. Such a waveform contains higher harmonics of frequency $3f_0$, $5f_0$ etc, but we may ignore these for the present purpose. Suppose the signal is a sinusoidal wave of frequency f. Then the result of combining the reference and signal voltages is a wave with frequency components $f + f_0$ and $f - f_0$ (see section 6.6). Since the filter only passes frequencies in the range 0 to f_b, the net result is that the lock-in amplifier plus the filter pass only those signal voltages whose frequencies lie in the range $f_0 \pm f_b$, or (70 ± 0.2) Hz. In other words the combination acts as a highly tuned circuit whose resonance frequency is locked to f_0.

Let us go back to the experiment. Without the lock-in amplifier the signal would be almost lost in the noise. However, the noise is distributed throughout the frequency spectrum, whereas the signal is entirely at a frequency of 70 Hz. Therefore, with the lock-in amplifier we retain the whole of the signal, but only that part of the noise that lies in the narrow frequency range $(70 \pm 0{\cdot}2)$ Hz. The noise is now much less than the signal. This technique for reducing the noise is known as *bandwidth narrowing.*

The technique is a powerful one and finds application in several branches of physics, for example in the measurement of light intensities. Fig. 7.12 shows a schematic arrangement for measuring the transmission of light through a thin slab of material as a function of wavelength. A beam of monochromatic light is modulated or chopped by allowing it to impinge normally on a rotating slotted disc C. Fig. 7.12 shows a disc that chops the beam four times per rotation. The frequency of rotation, of the order of 100 Hz, is not critical, provided the modulation frequency f_0 is not near the mains frequency or its harmonics.

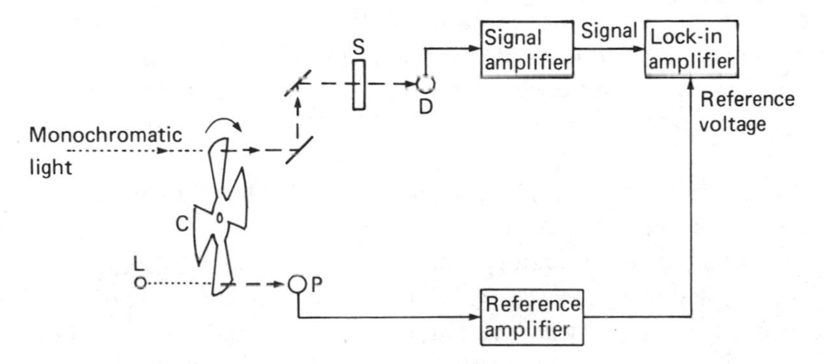

Fig. 7.12. Noise reduction in a measurement of light intensity. .... steady beam of light; ---- chopped beam.

The modulated beam passes through the sample S and is detected by a photodetector D, which produces an electrical signal that depends on the intensity of the light falling on it. The signal, a square-wave with frequency f_0, is amplified and passed into a lock-in amplifier. A light-emitting diode L provides a secondary source of light, which is also chopped by the rotating disc and then detected by a phototransistor P. The output from P, after amplification, is applied as the reference voltage to the lock-in amplifier. You can see that the arrangement acts in the same way as in the resistance measurement. The sample signal and the reference voltage, being modulated by the same rotating disc, always have the same frequency. The effects of extraneous light and of noise in the detecting and amplifying components are much reduced.

(g) *Voltage standards.* We return to the resistivity experiment. The method gives the resistivity in absolute units. The digital voltmeter used to measure the output of the current generator and to calibrate the lock-in amplifier is a commercial instrument capable of giving readings with a precision of about 1 part in 10^5. Such instruments have an internal voltage reference, commonly based on a zener diode. This is a semiconductor diode with the property that when the reverse bias voltage exceeds a certain value (known as the zener voltage) reverse current flows and rises rapidly with increasing voltage. Thus, by arranging that the current through the zener is kept constant within certain limits, the voltage across it is fixed within much closer limits. For diodes used as voltage references, the zener voltage V_z is about 6 V.

For a given zener diode, V_z varies with temperature T. Some values for a particular diode are given in exercise 4.4, p. 46. They correspond to a temperature coefficient $\mathrm{d}V_z/\mathrm{d}T$ of a few $\mathrm{mV\,K^{-1}}$, which is too large for a precise reference device. However, other types of zener diode have lower temperature coefficients - about $10\ \mu\mathrm{V\,K^{-1}}$ - and are suitable for the purpose.

The reference voltage of a particular instrument may be determined absolutely by comparison with a voltage standard which is ultimately related to those in a laboratory such as the National Physical Laboratory in the United Kingdom or the National Bureau of Standards in the United States. At one time these standards were based on temperature-controlled Weston standard cells, but the present voltage standard is derived from the ac Josephson effect. In the presence of microwave radiation of frequency f, the voltage across a Josephson junction has the value

$$V = n\frac{h}{2e}f, \tag{7.8}$$

where n is an integer, h is the Planck constant, and e is the elementary charge. Thus the measurement of a frequency, which can be made very precisely, provides an absolute voltage standard. For further details of voltage references see Horowitz and Hill 1980, p. 192. For a discussion of the Josephson effect and its use in voltage standards see Petley 1971 and 1980, and Gallop 1982.

7.3 Rayleigh refractometer

(*a*) *Description of instrument.* The Rayleigh refractometer is an instrument devised to measure the refractive indices of gases and also small changes in the refractive indices of solids and liquids.

Monochromatic light from a vertical slit S (Fig. 7.13) is collimated by an achromatic lens L_1 and falls on two vertical slits S_1 and S_2. The two beams pass through tubes T_1 and T_2 of equal length t, lying in the same horizontal plane. The beams then recombine to form vertical interference fringes in the focal plane of the lens L_2. The fringes are viewed by a small cylindrical lens L_3.

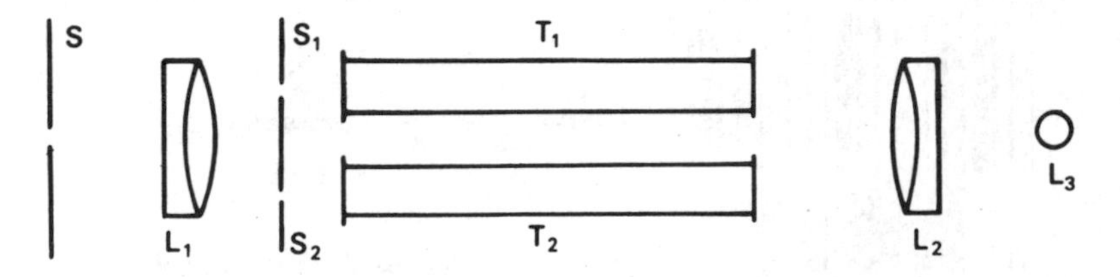

Fig. 7.13. Rayleigh refractometer – view from above.

To measure the refractive index of a gas, the fringes are first observed with both tubes evacuated. Gas is then admitted to one of the tubes, say T_1, thereby increasing the optical path of beam 1. This causes the fringe pattern to move sideways and the number of fringes passing a crosswire is counted. If p (not necessarily an integer) fringes pass, the refractive index μ of the gas is given by

$$t(\mu - 1) = p\lambda, \tag{7.9}$$

where λ is the wavelength of the light. We consider some practical aspects of the instrument in the following sections; a detailed account of the Rayleigh refractometer will be found in Candler 1951.

(*b*) *Reference system.* Instead of using a crosswire as the reference or fiduciary mark, it is better to use a second set of fringes which remain fixed. These are obtained by allowing only the upper halves of the beams from S_1 and S_2 to pass through the tubes. The lower halves pass under them and produce a second, independent set of fringes. Since the optical paths of the two lower beams are identical, the second set of fringes remains fixed throughout the experiment.

The advantage of this type of reference mark is that the eye is much more sensitive to relative displacements of two similar non-overlapping parallel lines, in this case the two sets of fringes, than to relative displacements between a crosswire and a fringe (see Fig. 7.14). In the first case, displacements of as little as $\frac{1}{40}$ of a fringe separation can be detected; whereas in the second, the limit is about $\frac{1}{10}$ of a fringe separation. The sensitivity of the eye in the first case is known as *vernier acuity*.

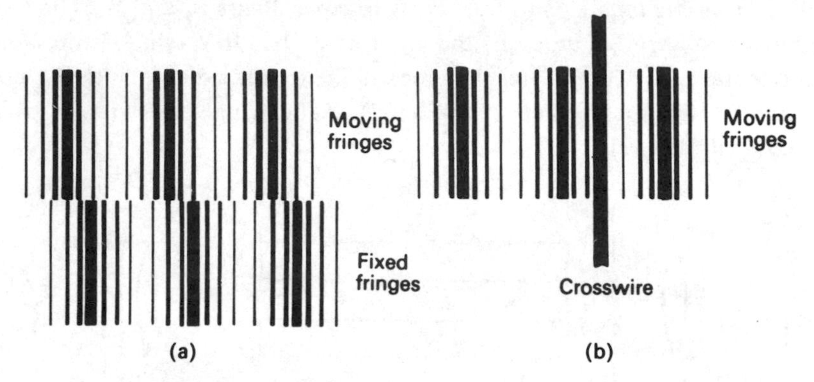

Fig. 7.14. Two forms of reference marks. (a) fixed set of fringes, (b) crosswire. The eye detects small movements more readily in (a) than in (b).

Another important advantage of the fringe reference system is that distortion of the framework of the apparatus or displacement of the double slit S_1S_2 does not affect the readings, because these faults affect both sets of fringes in the same way.

(c) Cylindrical eyepiece. In a practical case the separation s of the slits S_1 and S_2 is about 10 mm. The angular separation of the fringes is given by

$$s\theta = \lambda. \tag{7.10}$$

So for $\lambda = 500$ nm, θ is 5×10^{-5} rad or about $\frac{1}{5}'$. The fringes are thus very close together. They are viewed with the lens L_3, which is simply an accurately made cylindrical glass rod of diameter about 2 mm. The point of having a vertical cylindrical lens is that it only gives magnification in the direction of the separation of the fringes, which is where we want it. The effective aperture of the beam is much smaller than the pupil of the eye; any magnification therefore reduces the brightness of the field. If the magnification factor of the cylindrical lens is n (usually about 150), the brightness is reduced by a factor n. A spherical lens of the same magnification would reduce the brightness by a factor n^2. Since lack of brightness is one of the disadvantages of the instrument, the saving of the cylindrical lens is well worth having.

(d) Compensator method. It is tedious to count the large number of fringes that pass the crosswire as the gas is admitted into the tube T_1. The increase in the optical path length of beam 1 is therefore compensated by increasing the optical path length of beam 2 by a measured amount. This is another example of the null method. The two fringe systems are used as an indicator to determine when the two optical paths are equal.

Several methods of compensation have been devised. We shall confine ourselves to one. A fairly thin glass plate is placed in each of the two upper beams; one is fixed and the other can be rotated. When the two plates are parallel the path difference is zero. The path difference is varied by rotating the rod R (Fig. 7.15) to which the moving plate is attached. This is done by means of a micrometer screw M which bears against the radial arm A. In this way small changes in path difference can be made precisely and reproducibly.

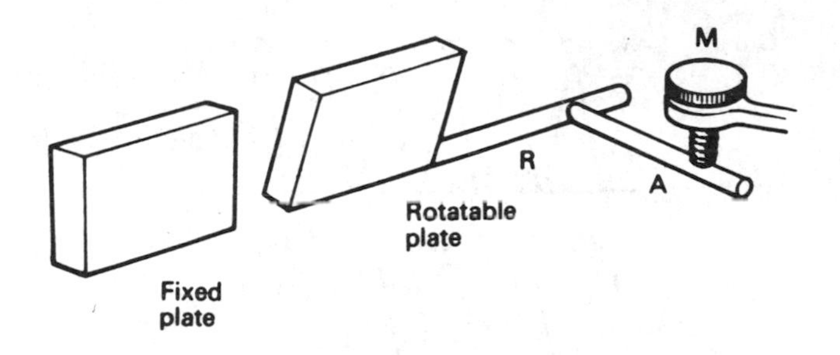

Fig. 7.15. Compensating device.

The path difference produced by the compensating device can be calculated as a function of the setting of the micrometer screw, provided we know the dimensions and refractive index of the rotating plate. But it is a good general principle to prefer an *empirical* to a calculated calibration. So the compensator is calibrated by observations of the fringe system with monochromatic light. The micrometer screw is rotated and its readings for displacement of 1, 2, 3, . . . fringes are noted. The calibration must be done at various wavelengths of incident light.

(e) White light fringes. All the discussion so far has been based on the use of monochromatic light. Clearly the ultimate measurements must be done in this way, because the wavelength is a crucial quantity in (7.9). However, it would be highly inconvenient if we used only monochromatic light, because all the fringes look alike. It would be necessary to sit watching the fringes as the gas was admitted into tube T_1 and to turn the screw on the compensating device at just the correct rate to ensure that the fringe system remained unchanged at all times. If the compensation was not exactly right and we lost sight of the fringes for an instant, we would not know which was the zero order fringe.

Fortunately this difficulty can be overcome by the use of a source of white light. Each monochromatic component in the source produces its own fringe system with its own spacing, the blue with the narrowest and the red with the widest (Fig. 7.16). What we actually see is the sum of all these fringe systems. For zero path difference, all the systems give a bright fringe, and the sum is a bright white fringe. But as the path difference increases, the systems get out of step and never get back in step again. (A *finite* set of different wavelengths may produce fringe systems that get back into step, but not an *infinite* set of continuously

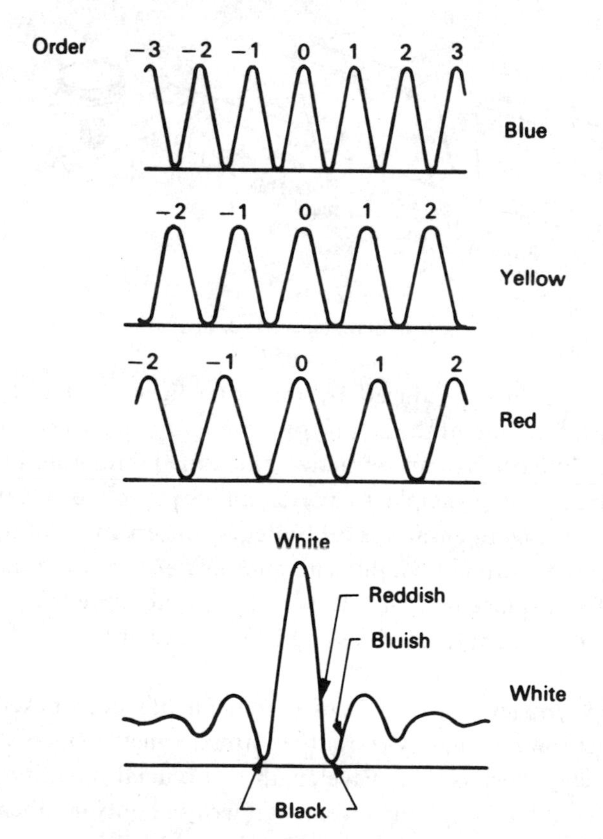

Fig. 7.16. Fringes for blue, yellow, red, and white light. The central white fringe is at zero order.

varying wavelengths.) The total effect is that the fringes on either side of the central white one are coloured. The colouring increases as we go farther out, and the fringes rapidly disappear. So with a white light source we have a means of detecting zero path difference.

The procedure in the experiment is therefore as follows. Wc start with both tubes evacuated and a white light source. The micrometer screw is set approximately so that both sets of fringes coincide. The white light is replaced by a monochromatic source, and the screw is set accurately. The required amount of gas is let into one of the tubes. The screw is then turned to bring the zero order fringe back into the central position. The fringe is located by means of white light as before, and the final adjustment made with monochromatic light.

(f) Dispersion effect. A difficulty arises over the white light method if the dispersion of the gas differs from that of the compensating plate.

Consider the situation when there is no gas in the tube and the compensator is set to zero. The fringe system with the white light source may be represented by the analytic diagram shown in Fig. 7.17a. This diagram is to be interpreted in the sense that if we draw a horizontal line across it, the crossing points give the fringe spacing at a particular wavelength. The lower we draw the line, the longer is the corresponding wavelength. Thus a horizontal line at the top gives the fringe spacing for blue light and at the bottom for red light.

The total effect of the white light is obtained by collapsing the diagram in the vertical direction. The zero order fringe, being originally vertical, collapses to a sharp dot, representing the central bright fringe, but the 1st, 2nd, 3rd, ... order fringes collapse to spread out lines which soon overlap, representing fringes that are blurred out.

Now suppose that the gas is admitted and the rotating plate set to compensate for the increase in optical path length. If the dispersions of the gas and the plate are the same, we can achieve compensation at all wavelengths, and the analytic diagram looks identical to Fig. 7.17a. But if the dispersions are different, we cannot achieve simultaneous compensation at all wavelengths. Suppose we set the plate to achieve exact compensation at some wavelength in the middle of the spectrum. The diagram would then look like Fig. 7.17b. The blue, say, is over-compensated and its zero order fringe has moved to the right; the red is under-compensated and its zero order fringe has moved to the left. If now we collapse this diagram vertically, it is fringe number 2 that becomes a dot. In other words, the white light fringe is no longer the one of zero

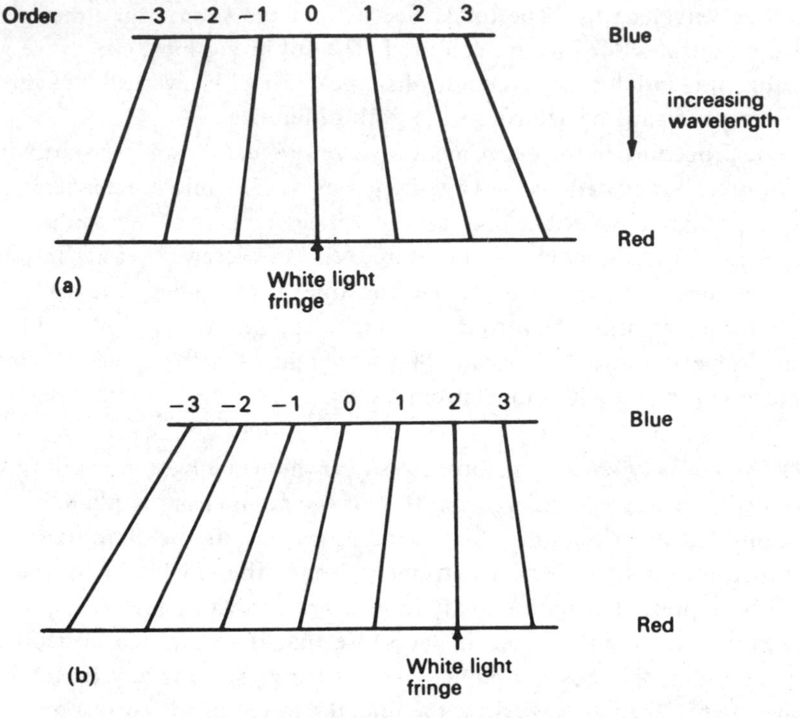

Fig. 7.17. Analytic diagram to show fringe system with white light. A horizontal line gives the fringe spacing at a particular wavelength. Collapsing the diagram vertically gives the result for white light. (a) Dispersions of gas and compensating plate the same, (b) dispersions different.

order. (In general none of the fringes in Fig. 7.17b is vertical, and the white light fringe is not of integral order.)

This defect in the ability of the white light fringes to pick out the zero order fringe may be overcome in two ways. The first is to admit the gas gradually so that we can follow the change in the white fringe. The other is to use a plate whose dispersion is close to that of the material under investigation. This is a technical matter which we shall not pursue here – see Candler 1951, p. 480 for further discussion.

(g) Precision and applications. We have already said that the smallest path difference that can be detected is about $\frac{1}{40}\lambda$. The largest that can be measured is about 250λ. We have

$$t\Delta\mu = p\lambda. \tag{7.11}$$

So for a tube 10 mm in length and $\lambda = 400$ nm, a change in μ of 10^{-6}

can be detected, and the maximum change that can be measured is 10^{-2}. For a tube 1 m in length, the smallest detectable change is 10^{-8}, and the maximum change is 10^{-4}.

The Rayleigh refractometer is the most precise instrument we have for measuring small changes in refractive index. The refractive index of a transparent mixture - liquid or gas - depends on the proportions of its components, and the refractometer is often the most precise device for determining the proportions or small changes in them. The instrument finds a variety of applications in this way in physics, chemistry, and biology.

For precise work the temperature must be strictly controlled - to obtain a precision of 10^{-6} for a liquid the temperature must be known to 10^{-2} K. The longer the tube, the more sensitive the apparatus, and the more difficult to obtain reliable results. Therefore we choose the shortest tube that will achieve the required precision in a given situation.

The refractometer was originally designed by Lord Rayleigh in 1896 to measure the refractive indices of the recently discovered inert gases, helium and argon. The value of $\mu - 1$ at s.t.p. is $3{\cdot}6 \times 10^{-5}$ for helium and $28{\cdot}1 \times 10^{-5}$ for argon.* As we have seen, the instrument is well able to measure even these low values.

7.4 Absolute measurement of the acceleration due to the Earth's gravity

(*a*) *Introduction*. Our fourth and final experiment is a precision measurement of *g*, the acceleration due to the Earth's gravity. Although for many purposes we only require the *variation* of *g* from place to place (see p. 126), there are some applications for which we need its *absolute* value. These arise primarily from the need to establish the unit of force in terms of the fundamental units of mass, length, and time. We can measure the mass of a suspended object in terms of the kilogram unit (p. 195). If we know the absolute value of *g* at the position of the object, we know the absolute value of the force that the object exerts on its suspension. Apart from practical applications, such as the absolute measurement of pressure and so on, we need the unit of force to establish the unit of electric current, since this is defined in terms of the force between current-carrying conductors. Absolute values of *g* are also required for astronomical purposes, such as the calculation of the motion of bodies in the solar system, including nowadays artificial satellites.

* Kaye and Laby, 1986. The values are for $\lambda = 589$ nm.

Until the Second World War the most precise method of measuring g was by the use of a reversible pendulum - basically the same as that used by Kater over a century before. However, the effects of small irregularities in the pendulum and its supporting knife-edge placed an ultimate limit of about 1 in 10^6 on the precision that could be obtained.*

The present method is to time the motion of a body in free fall and takes advantage of the very high precision with which small distances and short times can now be measured. The instruments based on this method are of two types - those in which a body is thrown up and measurements are made on the up and down parts of the motion, and those in which the body is simply allowed to fall. Each type has its advantages and disadvantages. The up-and-down instruments are less sensitive to air resistance effects, but care is necessary to eliminate vibrational effects from the body-launching system. The instrument to be described, by Zumberge, Rinker and Faller 1982, is of the second type. It has been selected because it contains a number of elegant and instructive features, and demonstrates the very high precision that can be obtained in an experiment, with ingenuity and care.

(b) Description of the method. The apparatus is shown schematically in Fig. 7.18. The dropped object C_1 is a cube-corner prism, known as a *retroreflector*, because it has the property that a ray of light, after internal reflection at each of three mutually perpendicular faces, returns travelling in the opposite direction. The principle is illustrated in Fig. 7.19 for the simple two-dimensional case.

C_1, and a similar cube-corner C_2, act as the ends of an arm of a Michelson interferometer. The light source for the interferometer is a specially stabilized He–Ne laser of wavelength λ.† The incident beam is split by a prism, and, after reflection at C_1 and C_2, the two beams are combined by the prism to give an interference pattern. A small part of the fringe pattern falls on a photodetector which gives an electric signal that depends on the intensity of the light.

During the measurements, C_2 remains at rest, while C_1 falls freely under gravity. This causes the fringes to move, and the output from the photodetector oscillates with a period that corresponds to a change in height of $\lambda/2$ for C_1.

The total distance that C_1 drops during the measurements is about 170 mm. The value of λ for the He–Ne laser is about 633 nm. The total

* See Cook 1967 for a good introductory article on the absolute determination of g.

† The method of stabilizing the laser is described by Baer *et al.* 1980.

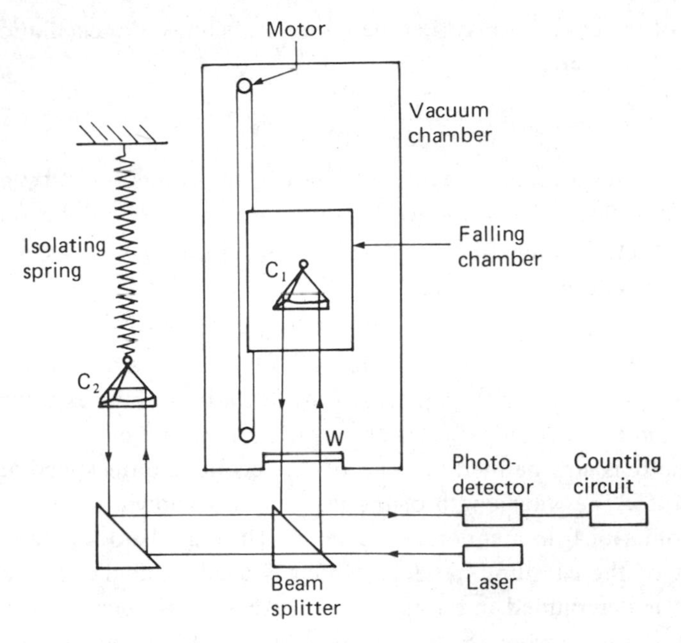

Fig. 7.18. Schematic arrangement of the apparatus of Zumberge, Rinker and Faller for the absolute measurement of *g*.

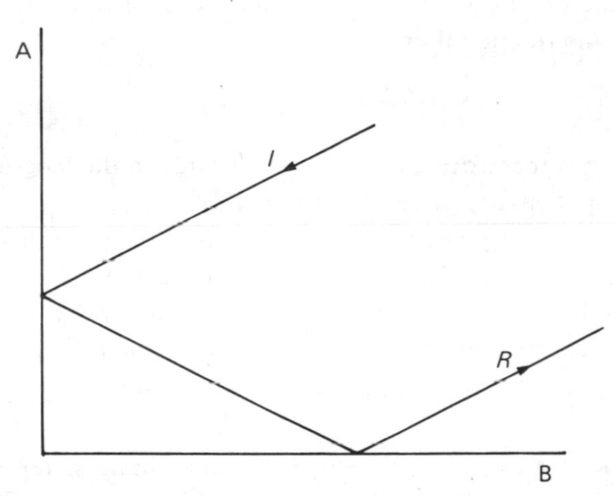

Fig. 7.19. Two-dimensional retroreflector. A and B are two mirrors mounted at right angles. An incident ray *I*, travelling in the plane perpendicular to the common axis of the mirrors emerges after two reflections as the ray *R* travelling in the direction opposite to *I*.

number of fringes that pass (i.e. the number of sinusoidal oscillations of the output) is therefore

$$N = \frac{170 \times 10^{-3}}{\frac{1}{2} \times 633 \times 10^{-9}} \approx 540\,000. \tag{7.12}$$

The time of occurrence of every 12 000th fringe is measured by means of a digital clock. Thus about 45 points are obtained in the relation between height h and time t as the body C_1 drops. These are fitted to the quadratic relation

$$h = ut + \tfrac{1}{2}gt^2 \tag{7.13}$$

by the method of least squares. This calculates the values of u and g which makes the sum of the squares of the deviations of the experimental values from the theoretical relation in (7.13) a minimum.

The metre is now defined in terms of the second and the speed of light (see p. 195). The wavelength of the laser is thus known in terms of the metre from a measurement of its frequency. The digital clock is calibrated in terms of the caesium standard which is used to define the second. Hence g is determined in terms of the metre and the second.

Let us consider what precision might be possible in the experiment. The photodetector can detect a change of about 2×10^{-3} of a fringe movement. So the fractional error in the length measurement is

$$\frac{\Delta h}{h} = \frac{\Delta N}{N} = \frac{2 \times 10^{-3}}{0{\cdot}54 \times 10^{6}} \approx 4 \times 10^{-9}. \tag{7.14}$$

The total time of fall is about 0·2 s, which can be measured to about 0·1 ns. Thus the fractional error in t is

$$\frac{\Delta t}{t} = \frac{0{\cdot}1 \times 10^{-9}}{0{\cdot}2} \approx 5 \times 10^{-10}. \tag{7.15}$$

The error in g is therefore governed by the error in the length measurement, and is potentially about 4 parts in 10^9. Since g is about 10 m s^{-2}, this corresponds to $\Delta g = 40$ nm s^{-2}.* However, to achieve this accuracy, or something close to it, the systematic errors must be reduced to nearly the same level, and this objective forms the basis of the design of the experiment. We shall now describe the features of the apparatus that reduce the systematic errors to the required levels.

(c) *Falling chamber.* It is clearly essential to eliminate as far as possible all non-gravitational forces on the dropped object. The two most impor-

* We continue to use SI units. However, a unit of acceleration commonly used in gravitational work is the gal. 1 Gal = 10^{-2} m s^{-2}, so 1 μGal = 10 nm s^{-2}.

tant here are air resistance and electrostatic forces. Air resistance is reduced by reducing the pressure in the chamber in which the object falls. However, at very low pressures, electric charge, which tends to build up on the object, cannot leak away, thus giving rise to electrostatic forces. The experimenters overcame this difficulty by allowing the object to fall inside a chamber which was itself falling.

The mechanism is shown in Fig. 7.20. Light from a light-emitting diode L is focussed by a sphere S attached to the cube-corner C_1 and falls on a detector. This is a device with two elements D_1 and D_2, each of which produces an electric current which depends on the intensity of the light falling on it. Both the light-emitting diode and the detector are fixed to

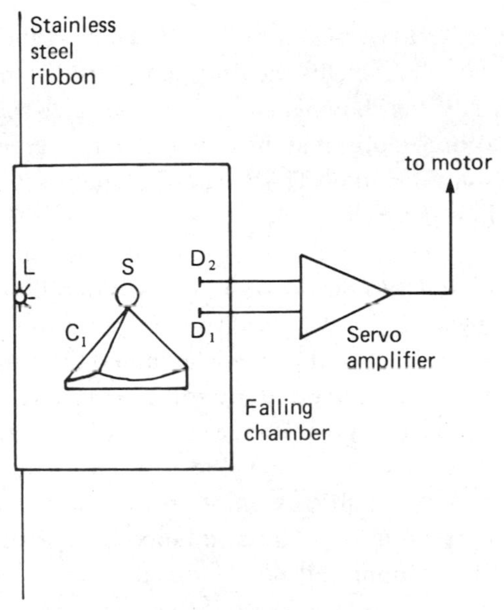

Fig. 7.20. Servo control of falling chamber. The motor, controlled by the output from the servo amplifier, drives the stainless steel ribbon to which the falling chamber is attached.

the chamber. If the dropped object C_1 falls slightly faster than the chamber, light from L falls on the element D_1; if C_1 falls slightly slower than the chamber, the light falls on the element D_2. The difference in the current outputs from D_1 and D_2 is amplified and made to drive a motor which controls the rate of descent of the chamber. Output from D_1 increases the acceleration of the chamber; output from D_2 decreases it. Thus the whole device is a servo system (see section 6.8) for ensuring that the chamber falls with the dropped object C_1. Notice that C_1 is falling freely and is unaffected by the action of the servo system, which acts only on the chamber. The chamber is vented to the vacuum chamber surrounding it, so that the air pressure in the two chambers is the same.

The falling chamber device has three main advantages. Firstly, it avoids the need for very low pressure to reduce air resistance. The dropped object and the air molecules in the chamber are all falling freely together. The absence of *relative* motion means that there is almost no force from the air molecules. In fact the experimenters still had to work at a reduced pressure (about 10^{-5} mm Hg). This was not due to air resistance, but to the effects of pressure and temperature gradients across the dropped object, which though small are nevertheless significant at the very high precision of the experiment.

The second advantage of the falling chamber is that, being made of conducting material, it shields the dropped object from external electrostatic fields. Lastly, the chamber can be controlled so that it gently arrests the fall of the dropped object at the end of the measurements and returns it quickly to its starting point. This means that many sets of measurements can be made in rapid succession.

(d) Long-period isolator. So far we have considered only the motion of the dropped object C_1, but we need to consider also the cube-corner C_2 which acts as a reference and which is assumed to be at rest. How is this ensured? C_2 must be supported in some way, and its support must ultimately be related to a point on the surface of the Earth. But this point will in general suffer acceleration due to both man-made and seismic vibrations. It is a reasonable assumption that these vibrations have no coherent phase relation to the times of successive drops of C_1. In other words, they give random and not systematic errors. However, there is no point in striving to reduce the systematic errors in an experiment if there remain substantial random errors. Suppose, for example, that for a single drop of C_1 the error in g due to the vibrations in the support of C_2 is 100 parts in 10^9, and that it is desired to reduce this error to 3 parts in 10^9. Since the error in the mean is $1/\sqrt{n}$ times the error in a single reading, the number of drops required to reduce the error to the desired value is about 1000. It is obviously preferable to reduce the error in a single reading, rather than to rely on the inefficient effect of the $1/\sqrt{n}$ factor, and the experimenters achieved this by means of a servo-controlled spring.

Suppose that the cube-corner C_2 is suspended from a spring to form an oscillating system of natural frequency f_0. If a sinusoidal motion of constant amplitude and variable frequency f is applied to the upper end of the spring, it is readily shown that, when the damping is small, the amplitude of the displacement of C_2 for the forced oscillations is approximately proportional to $1/(f_0^2-f^2)$. (When $f=f_0$ this expression becomes infinite, but in practice the damping force keeps the amplitude finite.) The important point in the present application is that, for $f \gg f_0$, the amplitude of the motion of C_2 becomes small, i.e. C_2 is effectively isolated from the motion of the upper end of the spring. For seismic vibrations the period is typically about 6 s. To isolate C_2 it is therefore necessary that the period of free oscillation of the spring should be about 60 s. This would require a spring with a length of about 1 km, which is clearly impracticable.

However, the experimenters were able to obtain an effective period of 60 s from a much shorter spring by electronic means. The principle of the method is as follows. Suppose we have a long spring of length L, fixed at its upper end, with a body attached to the lower end Q - Fig. 7.21. As the body makes vertical oscillations with amplitude X, a point

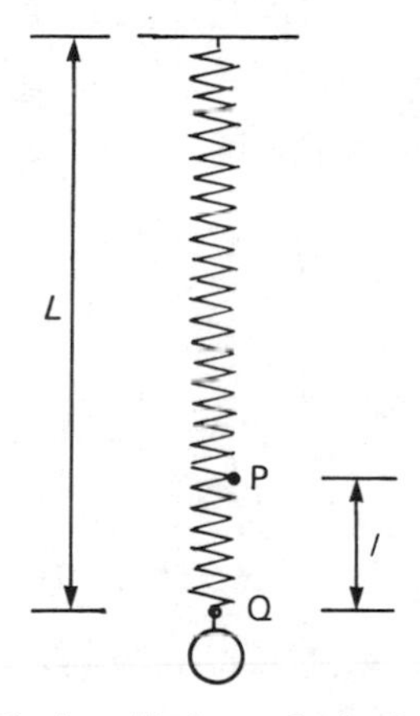

Fig. 7.21. Vertical oscillations of a body on a spring.

on the spring oscillates with amplitude proportional to its distance from the point of suspension. Thus a point P, distance l from Q, has amplitude $(1-l/L)X$. Now suppose we have a second spring of length l of the same material and with the same body suspended at its lower end. If, as the body oscillates with amplitude X, the upper end is made to oscillate with amplitude $(1-l/L)X$, the whole motion of the spring is the same as the lower part PQ of the first spring, and in particular the frequency of the motion is the same as that of the first spring.

The upper end of the spring is given the required motion by a servo system very similar to the one controlling the falling chamber. The upper end of the spring is attached to a cylindrical housing H – Fig. 7.22. Fixed to the bottom of the housing is a light-emitting diode L and a photodetector with two sensing elements D_1 and D_2. As before the sphere S attached to the cube-corner C_2 acts as a lens, and a signal arrives at the amplifier

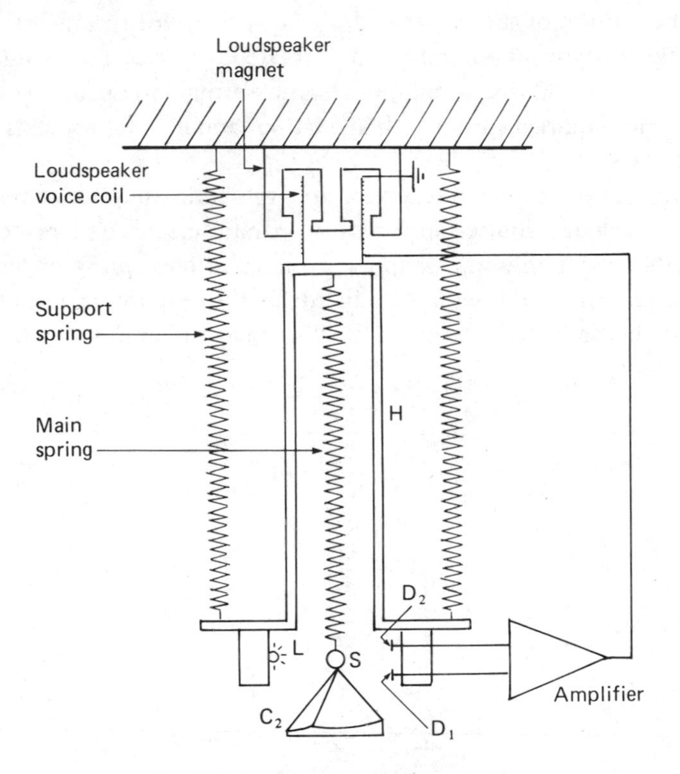

Fig. 7.22. Long-period isolation device.

that depends on the displacement of C_2 relative to the housing H. The output from the amplifier is fed into the voice coil of a loudspeaker, to which the upper end of the housing is attached. The gain of the amplifier is adjusted to make the amplitude of the oscillation of the housing some required fraction of the amplitude of the body C_2. The closer the fraction is to unity, the larger is the period of oscillation of the spring. In this way the experimenters were able to use a spring of length 1 m whose period of oscillation corresponded to a spring of length 1 km.

(e) Other errors. The experimenters took many other precautions to reduce systematic errors, of which we shall mention only two. The first concerns the optical path. Clearly the light beams must be vertical; otherwise the length measurements will not give the height h, but $h \cos \phi$, where ϕ is the angle between the light beams and the vertical. There is a subtle effect which needs to be considered in this connection. The window in the vacuum chamber (W in Fig. 7.18) is tilted slightly in order to avoid reflection of the light back to the laser. This by itself does not cause a deviation of the beam from the vertical, because the two surfaces of the window are parallel. However, the difference in air pressure on the two sides means that the refractive indices of the air on the two sides are slightly different, and this, together with the tilt, does give a deviation. Fortunately, the effect is negligible, provided the angle of incidence of the light is close to zero.

A second type of error arises from the electronic circuits that amplify the oscillating electric signal from the detector registering the passage of the interference fringes. As the dropped object descends, the frequency of the signal increases linearly with time, from about 1 MHz at the beginning of the measurements to about 6 MHz at the end. The amplifying circuits produce phase shifts which depend on frequency. It may be shown that phase shifts that vary linearly with frequency do not affect the results - only a non-linear variation gives rise to error. The phase changes depend on the bandwidth of the amplifier.* The experimenters found that a bandwidth from zero to about 30 MHz was necessary to reduce the error to an acceptable level.

(f) Results. The apparatus has been tested in several places, including Boulder, Colorado, and Sèvres, near Paris. An example of a series of measurements made at Boulder in May 1981 is shown in Fig. 7.23. The results show the variation of g with time, at the same place, due to the tidal redistribution of water in the oceans. This variation has several components with different periods, the two most prominent having periods of a day, and half a day.† They are clearly seen in the figure. Each experimental point is the result of a set of 150 drops. The rms deviation of the points after the variation of the tides has been removed is 60 nm s^{-2}.

Measurements taken at different places indicate an average error of about 100 nm s^{-2}, but the experimenters hope to reduce this to about

* See Millman 1979, chapter 13.

† See Cook 1973, chapter 4, for the theory of the variation of g due to the tides.

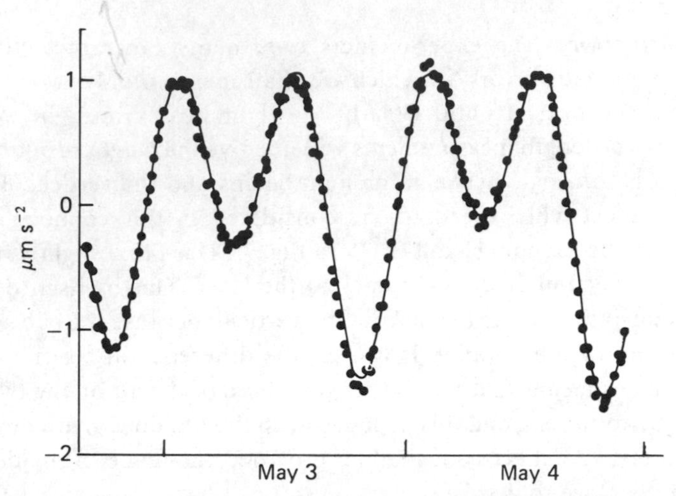

Fig. 7.23. The tidal variation of g at Boulder, Colorado measured with the apparatus of Zumberge, Rinker and Faller in May 1981. The curve shows the theoretical variation.

60 nm s^{-2}. It may be noted that this error is about 150 times smaller than that of the most accurate reversible pendulum. A striking illustration of the phenomenal precision of the apparatus is that its error is equivalent to the change in g due to a change in height of 20 mm (!) at the surface of the Earth.

Exercises

7.1 It is required to pass current into a potentiometer of resistance R connected across JK in Fig. 7.5, a fraction u of the current to come from E and the remainder from G. Show that if the emf of G is v times that of E, then S must be chosen to satisfy

$$\frac{S}{R}=\frac{v-1}{1-u}.$$

7.2 Prove the result in (7.5).

7.3 A zener diode with $V_z = 6$ V operated at a current $I_z = 1$ mA has a dynamic resistance $\mathrm{d}V_z/\mathrm{d}I_z = 3\ \Omega$. If the current varies by 2%, what is the fractional change in the voltage?

7.4 Estimate the increase in height at the surface of the Earth corresponding to a reduction in g of 6 parts in 10^9.

7.5 In the present chapter four pieces of apparatus are described and analysed. Treat the Fabry–Pérot interferometer in the same way, including in the discussion the method of exact fractions.

7.6 Explain the doublet method used in precision mass spectroscopy.

8

Experimental logic

8.1 Introduction

Systematic error is just a euphemism for experimental mistake. Such mistakes are broadly due to three causes:

(a) inaccurate instruments,
(b) apparatus that differs from some assumed form,
(c) incorrect theory, that is, the presence of effects not taken into account.

We have seen the remedy for the first - calibrate. There is no blanket remedy for the other two. The more physics you know, the more experience you have had, the more likely you are to spot the effects and hence be able to eliminate them. However, there are ways of making measurements, of following certain sequences of measurements, which automatically reveal - and sometimes eliminate - certain types of error. Such procedures form the subject of the present chapter. Some are specific, others are more general and add up to an attitude of mind.

Finding and eliminating a systematic error may sound a negative, albeit desirable, object. But there is more to it than that. The systematic error that is revealed may be due to a phenomenon previously unknown. It is then promoted from an 'error' to an 'effect'. In other words, by careful measurement we may make discoveries and increase our understanding of the physical world.

8.2 Apparent symmetry in apparatus

It is a good rule that whenever there is an apparent symmetry in the apparatus, so that reversing some quantity or interchanging two components should have no effect (or a predictable effect - see the third example), you should go ahead and make the change. A few examples will illustrate the point.

(a) Suppose we are comparing two resistances by means of a potentiometer (see section 7.1) and know nothing about the thermoelectric effect. The apparent symmetry element in the apparatus is the direction of the current in the two circuits. It appears that the balance point should be the same if we reverse both currents. We do so and find it is not. The operation has revealed that something is happening that we had not reckoned with. We investigate the phenomenon and find it is due to a thermoelectric emf, which is independent of the direction of the two main currents. In this case, simply averaging the two balance points eliminates the error.

(b) Consider an experiment, for example, the measurement of the thermal conductivity of a material, in which we need to measure the temperature difference $\Delta\theta$ between two points P and Q. Suppose we do this by measuring the temperature at P and Q with a pair of similar thermometers. Symmetry says that interchanging the two thermometers should not affect the result. We interchange them and find that it does, thereby discovering that the thermometers are not reading correctly. If $\Delta\theta$ is small, its value from one pair of temperature readings could be seriously wrong. Interchanging the thermometers and taking the mean of the two values of $\Delta\theta$ considerably reduces the error. (If $\Delta\theta$ is small, a better method still is to measure it directly and thus avoid the unsatisfactory procedure of taking the difference between two nearly equal quantities. This could be done by placing platinum resistance thermometers at P and Q and connecting them in the opposite arms of a Wheatstone bridge.)

(c) The third example is the Wheatstone bridge. Look at the circuit in Fig. 8.1. R is an unknown resistance, and S is a standard of known value. The value of R is obtained by finding the value of AB at balance. Denoting it by x_1 and the value of AC by l, we have

$$\frac{R}{S}=\frac{x_1}{l-x_1}. \tag{8.1}$$

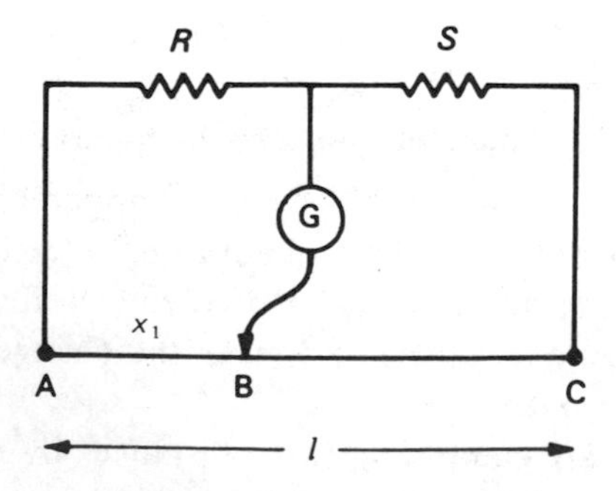

Fig. 8.1. Wheatstone bridge.

Symmetry says that if we interchange R and S, the new value of AB should be

$$x_2 = l - x_1. \tag{8.2}$$

We make the change and get a different value. The operation has revealed the presence of end effects. In this case, replacing x_1 in (8.1) by the mean of x_1 and $l - x_2$ does not eliminate the error.

8.3 Sequence of measurements

The *order* in which measurements are made can be very important, as the following example illustrates. Three students are asked to find how the terminal velocity of a sphere falling in a liquid varies with its diameter. They are given a set of 4 ball bearings of various sizes and a large tank of glycerine.

Student X takes the largest ball and measures its terminal velocity five times, then takes the next largest and does the same, and carries on until he reaches the smallest. He gets a poor result. Why? Because the laboratory has been warming up during the measurements, and so has the glycerine. The viscosity of glycerine, like that of most liquids, drops rapidly with increase in temperature. The terminal velocity depends on the viscosity of the liquid as well as the diameter of the ball. Therefore, since the average viscosity was different for each ball, the results do not give the variation of terminal velocity with diameter alone.

Student Y knows more physics than X and knows about the variation of terminal velocity with viscosity and hence with temperature. He therefore builds a device for keeping the temperature of the glycerine constant. He makes the measurements in the same sequence as X and gets a much better result, but it is still incorrect. Why? Because unknown to Y the clock that he is using to measure the terminal velocity is gradually slowing down, and this gives an effect systematically different for the 4 balls.

The third student Z is as ignorant as X about the effect of temperature on the measurements, and his clock is just as poor as Y's, but he gets a very good result. This is due to the sequence in which he – instinctively – makes his measurements.

Denote the 4 balls by A, B, C, D. Suppose that instead of 5 successive measurements for ball A, followed by 5 for ball B and so on, the measurements are made in the order

ABCDABCD . . .

Now instead of ball A being measured with the high and D with the low viscosity liquid, all 4 balls are being measured at a high value and then again at a lower value and so on. However, although this sequence reduces the systematic error considerably, it is still true that for each set of 4 measurements A comes at the high and D at the low viscosity end. So even better is the sequence

ABCDDCBA,

which is repeated as many times as the total time permits. This is the way student Z makes his measurements. (An extra precaution would be to make the next sequence BCDAADCB and so on.) You can see that over the entire sequence of measurements, the effects of a smooth variation with time of the viscosity of the liquid, or of the accuracy of the clock or indeed of any factor other than the diameter of the ball, will probably be small.

Note that even Z's method can be improved upon. His ignorance of the temperature effect is hardly a merit. The measurements in this experiment are so sensitive to the temperature of the liquid, that a competent experimenter would not only adopt Z's sequence, but would also measure the temperature from time to time to check that there was no chance correlation between the temperature variation and the sequence of the diameters.

8.4 Intentional and unintentional changes

In an experiment to measure the effect of varying one quantity, we obviously try to keep all other quantities constant. However, there is always the possibility of variations in the latter, and in the last section we described a method for reducing the effects of such variations. The method is very effective, but is only applicable when the unwanted variations are not caused by or related to the quantity that we wish to vary. This is clearly the case in the previous example. Neither the temperature of the glycerine nor the accuracy of the clock depend on which ball we decide to drop for the next measurement.

However, consider the following experiment. We wish to investigate the change in the dimensions of a ferromagnetic material caused by the application of a magnetic field, a phenomenon known as *magnetostriction.* An iron rod is placed in a solenoid, and its length is measured as a function of the current through the solenoid, the current being a measure of the magnetic field.

Now the change of length in magnetostriction is small – the fractional change for complete magnetization is about 5×10^{-5} – and so in order to measure it accurately, we must keep the temperature of the sample constant; otherwise thermal expansion will swamp the magnetic effect. When we increase the current through the solenoid we increase the heat generated in it, and this may raise the temperature of the specimen. The method of the last section is quite irrelevant here - the quantity we are varying is *causing* the unwanted variation. What we have to do is to ensure that the current through the solenoid does *not* affect the temperature of the specimen, for example, by winding the solenoid on a water-cooled former.

The converse effect is also a potential source of error. The current through the heating coil of a furnace may give rise to a magnetic field, which in turn may affect the measurements.

8.5 Drift

In section 8.3 we had an example of slow systematic variation or *drift* during an experiment. Apart from temperature, other common quantities that are liable to vary are atmospheric pressure and humidity, the voltage of a battery, the mains voltage and even its frequency.

Choosing an appropriate sequence for the measurements is one way of reducing the effects of these variations, but often we wish to prevent or at least minimize the variations in the first place. This is usually done by various negative feedback and servo devices (sections 6.7 and 6.8).

Section 8.3 also provided an example of *instrumental* variation. We should always have it in mind that instruments are liable to drift - their zero errors may change and so may their sensitivities. It may therefore be necessary to calibrate an instrument more than once - indeed many times - during an experiment.

Notice that the calibration operation itself may form part of a sequence which can give rise to a systematic error. Suppose, for example, we are comparing two voltages V_1 and V_2 by means of a potentiometer. The emf of a potentiometer cell tends to fall with time. So if, after standardizing the instrument, we always measure V_1 first and then V_2, the measured value of V_1/V_2 will be systematically low.

8.6 Systematic variations

Look at the numbers in Table 8.1 which represent the measurements of the diameter d of a piece of wire at various points along its length x. If you were asked for the best value of the diameter and an estimate of the

Table 8.1. *Measured values of the diameter of a wire at various points along its length*

Length/m	Diameter/mm
0·0	1·259
0·0	1·263
0·0	1·259
0·0	1·261
0·0	1·258
0·1	1·252
0·2	1·234
0·3	1·209
0·4	1·214
0·5	1·225
0·6	1·248
0·7	1·258
0·8	1·256
0·9	1·233

standard error in a single measurement, how would you proceed? (Stop at this point and decide.)

Let us see how our friends X and Y would tackle this problem. X is in no doubt. He has been told that the best value of a quantity is the mean of a set of measurements, and he has a calculator that calculates the mean and the standard error. So he happily feeds all the readings into his calculator and finds the mean, which is 1·245 mm, and the standard error, which is 0·018 mm.

Y, however, notices that the readings are not varying in a random manner and so he plots them on a graph, which is shown in Fig. 8.2. It is now obvious that the variation is systematic. He realizes that the mean of all the readings has no significance at all. The diameter was measured five times at $x=0$, so its value there is over-weighted. Accordingly, he replaces these five readings by the single number 1·260, which is their mean. He takes the mean of the ten values which he now has and obtains 1·239 as his best value.

Furthermore, he realizes that, since the diameter is undoubtedly varying along the length of the wire, the spread in the values over the whole range of x has nothing whatever to do with the standard error in a single measurement. In order to obtain the latter, he looks at the spread in the five values at $x=0$ and obtains 0·002 mm as an estimate for σ. (Whether

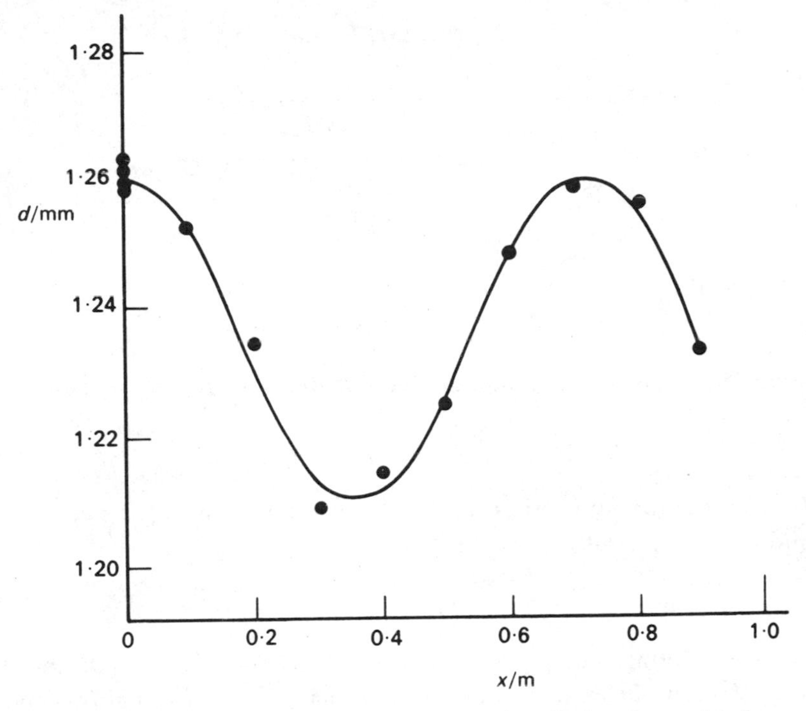

Fig. 8.2. Diameter of wire d at various points along its length x – plot of values in Table 8.1.

this is a random error or whether the cross-section is not circular at $x = 0$ we cannot say without being told more about the measurements.)

Y's approach is a sensible one, but there is a point to be noted in connection with the 'best value' of d. Since d is varying in a systematic way, the value we require is not necessarily d_m, the mean obtained by Y. If, for example, we have measured the resistance of the wire and wish to determine the resistivity of the material, the quantity required is the average value of $1/d^2$, which is not quite equal to $1/d_m^2$. In the present case the difference is small, but occasionally it is not, and the correct average must be taken.

Another situation that calls for examination is a set of results that spread by more than their errors indicate. Consider the set of results for the speed of sound in air at room temperature given in Table 8.2. We may suppose they were obtained by measuring the wavelength of stationary waves at various frequencies in a resonance tube. Suppose that at each frequency many measurements were made and that their internal

Table 8.2. *Measured values of the speed of sound*

Frequency/Hz	Speed/m s^{-1}
1000	346·7
720	341·5
200	338·6
600	342·2
380	339·6

consistency was such that the standard error in each result was estimated to be

$$\sigma = 0{\cdot}7 \text{ m s}^{-1}. \tag{8.3}$$

In this situation some students simply take the mean of the five results and give as its error

$$\sigma_m = \frac{0{\cdot}7}{\sqrt{5}} \approx 0{\cdot}3 \text{ m s}^{-1}, \tag{8.4}$$

blithely ignoring the fact that three of the results are 3σ, 4σ, and 7σ away from the mean. If the value of σ given in (8.3) is reasonably correct, this is clear evidence of some systematic effect, and until it is discovered neither the mean nor the value of σ_m can be considered of much significance.

With all wave motion there is the possibility that the speed varies with frequency, a phenomenon known as *dispersion.* For sound waves in air, careful measurements by many experimenters have shown that there is no measurable dispersion at the frequencies in Table 8.2. However, in any resonance tube experiment there are certain corrections to be made, one of which depends on frequency (see Wood 1940, chapter X). It is possible that a systematic error in this correction has caused the present variation. Alternatively there might be a systematic error in the frequency values.

We therefore plot the values of the speed against frequency - Fig. 8.3. There does appear to be a correlation, and, if it is easy to make the measurements, it would be worth making a few more at other frequencies to see if the trend were confirmed. If so, we should look very carefully at the frequency-dependent correction and also at how the frequency values were obtained.

If the trend were not confirmed, we would have to look elsewhere. For example, although the value of σ in (8.3) might be correctly calculated

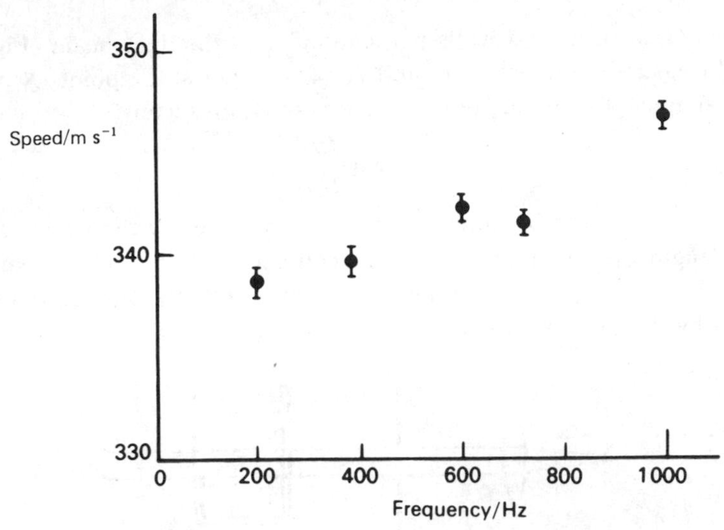

Fig. 8.3. Measured values of the speed of sound at various frequencies – plot of values in Table 8.2.

from the measurements, it might be the case that the spread in the values at each frequency is spuriously small. This could arise if the resonance condition were being detected by ear, and a succession of measurements were made at the same frequency. The experimenter might have been influenced by the first reading at each frequency, and tended to find subsequent resonance conditions close to the first. This could be avoided by making only one or two measurements at each frequency and then repeating this cycle a few times without looking at the previous results.

We have treated this example in some detail to show possible ways of proceeding when results spread more than their apparent error suggests. It is a not uncommon situation in experimental work.

8.7 Calculated and empirical corrections

In many experiments, corrections have to be made to take account of systematic effects. In estimating the magnitude of these corrections, preference should always be given to empirical methods, i.e., methods based on actual measurements, rather than to theoretical calculations. The latter may be wrong for a variety of reasons – wrong theory, incorrect assumptions, faulty calculations – whereas empirical methods are, by their very nature, less likely to be in error.

Suppose, for example, we are investigating the transmission of light of a given wavelength through a certain liquid. We place the liquid in a glass cell with its end walls perpendicular to the light beam (Fig. 8.4) and measure the intensity I_X and I_Y of the light at the points X and Y. We suppose for simplicity that the transmission factor

$$f = \frac{I_Y}{I_X} \tag{8.5}$$

has been found to be independent of I_X. We require the value of f for the length l of the liquid alone. The cell cannot be entirely transparent, so it is necessary to correct the measurements for the attenuation of the light by the two end walls.

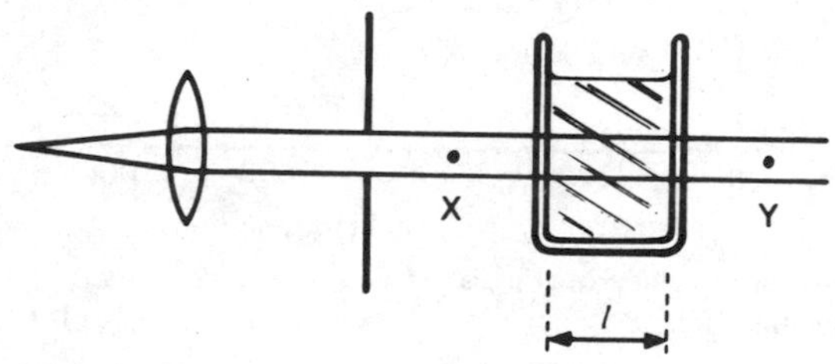

Fig. 8.4. Arrangement for measuring the attenuation of a light beam by a sample of liquid.

The theoretical way of making the correction would be to measure the thickness of the walls and look up in some table of physical constants the attenuation produced by this thickness of this particular glass for light of the wavelength we are using. Assuming that the information exists, the correction would depend on our knowing the correct wall thickness – it might not be constant and we would have to get the correct average value over just the area through which the light passed. It would depend on our knowing the wavelength of the light, and above all it would depend on our cell being made of just the same type of glass as that to which the tables referred.

The empirical method is first to measure the intensity at X and Y for the empty cell, then, without displacing the cell relative to the light beam, to fill the cell with the liquid and repeat the measurements. You can see that this procedure eliminates every one of the difficulties mentioned in the last paragraph.

Although empirical should be preferred to theoretical corrections, the best procedure of all is to obtain the corrections both ways and check that they agree. An example of this occurs in an experiment by Froome, who measured the speed of light using microwave radiation (p. 129). The only part of the experiment that concerns us here is the measurement of the wavelength λ of the radiation. The principle of the measurement is shown in Fig. 8.5. The microwave signal from the source S is divided in two and applied to two transmitting horns T_1 and T_2. The two signals are received by the horns R_1 and R_2 and combined in a unit not shown. The magnitude of the resultant signal depends on the relative phases of the signals received by R_1 and R_2. If they are exactly in phase, it is a maximum; if they are exactly out of phase, it is zero. (Whatever the relative magnitudes of x_1 and x_2, the signals received by R_1 and R_2 are made equal in amplitude. This is achieved by adjusting the relative amplitudes of the signals transmitted by T_1 and T_2 without altering their phases.)

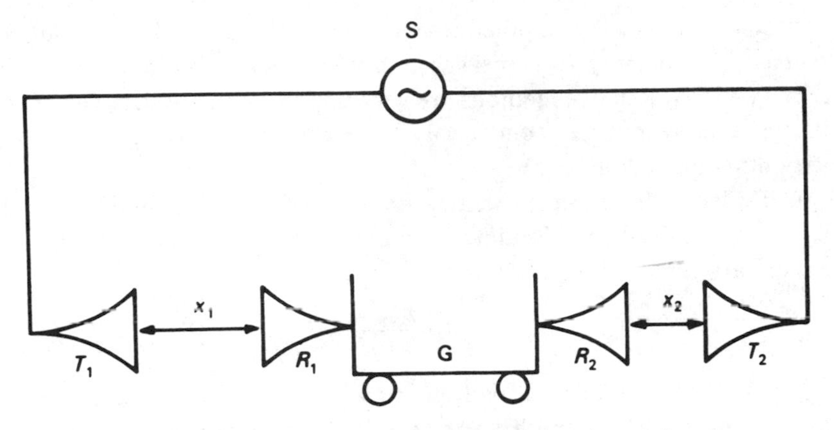

Fig. 8.5. Measurement of the wavelength in Froome's experiment.

R_1 and R_2 are mounted on a rigid trolley G so that x_1 and x_2 can be varied, the quantity x_1+x_2 remaining constant. If the trolley is moved so that x_1 increases by $\lambda/2$, then the phase of the signal at R_1 is retarded by π and at R_2 advanced by π. Thus the resultant signal goes through a complete cycle during the motion. The value of λ is obtained by counting the number of cycles through which the resultant signal passes as x_1 or x_2 is changed by a known amount.

Now the statement that increasing x_1 by $\lambda/2$ causes the phase of the signal at R_1 to be retarded by π, is only strictly true in the limiting case where T_1 is a point emitter and R_1 a point receiver. In practice, radiation is emitted and received over a finite area. The path lengths of the rays from the various parts of T_1 to the various parts of R_1 vary. Moreover, when the value of x_1 is changed, the different path lengths are changed by different amounts. The correction for these *diffraction* effects is important in a precision experiment such as this.

Froome's procedure was to make measurements with various types of covering of the horns and at various values of x_1+x_2. The diffraction effect, which varied, was calculated theoretically in each case. The agreement between theory and experiment throughout showed that in any given situation the correction was being calculated correctly and could be applied with confidence. This dual approach - theoretical and empirical - is the very acme of the experimental method.

8.8 Relative methods

In section 7.1 we described the measurement of the ratio R_1/R_2 of two resistances by means of a potentiometer. This is an example of a *relative method.* The quantity R_1 is measured, not absolutely, but in terms of, or relative to, R_2. Relative methods are very important in physics. They can be made more precisely and easily than absolute measurements, and very often are all we require.

Consider as an example the measurement of the viscosity of a liquid by a method based on Poiseuille's formula for the flow of a liquid through a capillary tube. The formula is

$$\frac{\mathrm{d}V}{\mathrm{d}t}=\frac{p\pi r^4}{8l\eta}, \tag{8.6}$$

where $\mathrm{d}V/\mathrm{d}t$ is the rate of volume flow of the liquid, p the pressure drop along a tube of length l, r the internal radius of the capillary tube and η the viscosity of the liquid. If we maintain and measure a constant pressure difference along the tube, and measure $\mathrm{d}V/\mathrm{d}t$, l and r, we can calculate the viscosity. This is an absolute measurement.

Now look at the apparatus in Fig. 8.6, which is a viscometer due to Ostwald. A fixed volume of liquid of density ρ_1 and viscosity η_1 is introduced into A and sucked into B so that its level is just above L on the left and N on the right. The time τ_1 taken by the liquid to drop from L to M - both levels being precisely marked - is measured. The liquid is then replaced by a second liquid of density ρ_2 and viscosity η_2, and the corresponding time τ_2 is measured. It is easy to show that the ratio of the viscosities is given by*

$$\frac{\eta_1}{\eta_2}=\frac{\rho_1\tau_1}{\rho_2\tau_2}. \tag{8.7}$$

* Equation (8.7), like Poiseuille's formula on which it is based, ignores the fact that the liquid acquires kinetic energy. A small correction is necessary to take account of this - see Smith 1960, chapter XI, for details.

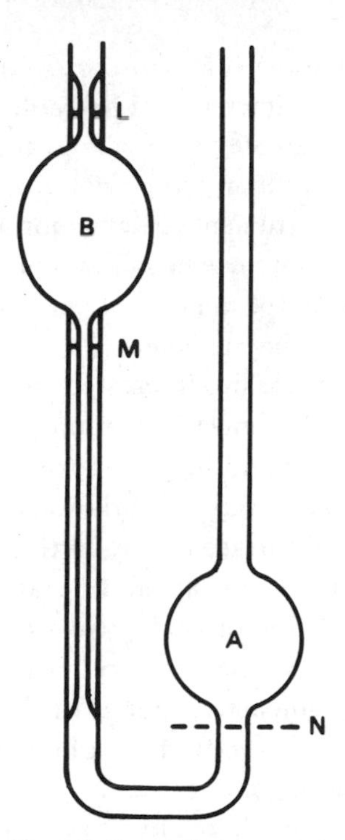

Fig. 8.6. Ostwald viscometer.

The measurement of the quantities on the right-hand side of (8.7) is comparatively simple.

Notice that the relative method completely avoids two difficulties in the absolute method. The first is maintaining and measuring a constant pressure head, and the second is measuring the internal radius of the capillary tube, which must be done precisely since r is raised to the fourth power in the formula. Furthermore, the simplicity of the Ostwald apparatus means that temperature control is easier. As mentioned earlier, temperature has a marked effect on viscosity; so this is an important advantage.

The relative methods described so far give the *ratio* of two quantities, but a relative method can also give the *difference* between two quantities. A good example of this occurs in the measurement of g, the acceleration due to gravity. We have seen (p. 103) that to measure the absolute value of g with high precision an elaborate apparatus is necessary. However, it is possible to obtain precise values for differences in g with much simpler apparatus. The instrument used for the purpose is a *spring gravimeter*, which is essentially a very sensitive spring balance. There are several versions for use under different conditions. (See Bomford 1980, p. 373 for details.) The gravimeter must be calibrated by observations at a few places where the absolute value of g is already known. But it can measure differences in g with an error smaller than that of the absolute calibration values. The instrument is light, portable, and convenient to use, so that many observations can be made rapidly. Modern gravimeters have sensitivities of about $100\,\mathrm{nm\,s^{-2}}$, which is comparable to the precision of the best absolute instruments.

For many purposes we are not interested in the absolute value of g, but only in its variation. The most important factors in the variation of g with position are latitude and height above sea-level - see Appendix G. When these have been allowed for, variations of g over large distances are related to the sea-level surface of the Earth and provide information about the structure of the continents and oceans; variations over small distances provide information of geological structure.

Other examples of relative methods are measurements of such quantities as the strength of a radioactive source, the intensity of a light source, and the flux density of radiation from a radio galaxy. In all these cases, the absolute determination of the quantity is very difficult, and the measurements are usually made relative to another similar quantity.

Notice finally that if we know the absolute value of the quantity we are measuring for *one* sample or object, all the relative values are converted into absolute values. Thus, once we have put one liquid of known viscosity and density through an Ostwald viscometer, measurements for another liquid give its viscosity absolutely.* And once the value of g is determined absolutely at one of the points on the network of difference values, all these values become absolute.

8.9 Why make precise measurements?

We said in chapter 2 that the precision to be sought in an experiment depends on its object. That is generally true, but in many experiments in physics, and particularly in measurements of fundamental quantities, we simply do not know what precision may, in the last resort, be appropriate. We try to attain the very highest precision that known phenomena and techniques permit. Why is this?

* The kinetic energy effect mentioned in the footnote on p. 124 requires that *two* liquids of known viscosity be put through the apparatus. But as you will already have realized - if you have caught on to the spirit of the last few chapters - the most satisfactory procedure is to calibrate the apparatus with *several* liquids of known viscosity.

If you look at the article entitled 'The 1986 Adjustment of the Fundamental Physical Constants' (Cohen and Taylor 1987), you will see that several of these constants are thought to be known to one part in a million or even better. Now you may wonder if there is any point in this, or whether making such measurements is a useless exercise like calculating the value of π to hundreds of decimal places.

The answer is quite simple. Precise experiments have a very important purpose. They test our theoretical ideas and, when they give results at variance with them, lead to new theories and discoveries. There have been many examples of this in physics and chemistry. A theory says that two quantities are equal. We do an experiment and find that, within the limits of this experiment, they are. We then do a more precise experiment and find a small difference. In other words the theory is only a first approximation. The more precise experiment guides us in the next theoretical step. We conclude with a few examples of discoveries that have been made as a result of careful and precise measurements.*

(a) Prior to 1894 it was thought that, apart from variable quantities of water vapour and traces of carbon dioxide, hydrogen, etc., atmospheric air consisted of oxygen and nitrogen. However, careful measurements by Rayleigh showed that the density of the gas remaining when the oxygen had been removed was about $\frac{1}{2}\%$ higher than the density of pure nitrogen obtained from a compound such as ammonia. This led Rayleigh and Ramsay (1895) to the discovery of the inert gas argon, which is now known to constitute about 1% of the atmosphere.

(b) The discovery of deuterium is another example of the fruitfulness of exact measurement. In 1929 the ratio of the mass of the hydrogen atom to the mass of ^{16}O (the isotope of oxygen with mass number 16) was found by chemical determination of atomic weights to be

$$\frac{\text{mass H}}{\text{mass }^{16}\text{O}} = \frac{1{\cdot}00799 \pm 0{\cdot}00002}{16}. \tag{8.8}$$

In 1927 Aston had measured the ratio in his mass spectrograph and found

$$\frac{\text{mass H}}{\text{mass }^{16}\text{O}} = \frac{1{\cdot}00778 \pm 0{\cdot}00005}{16}. \tag{8.9}$$

The discrepancy between these two values led Birge and Menzel (1931) to suggest that what was being measured in the chemical determination

* A detailed account of the measurement of the fundamental constants has been given by Petley (1985). An earlier article by Cohen and DuMond (1965) is well worth reading for its discussion and comments on the experimental method. For an instructive article on the importance of precise measurements in physics, see Cook (1975).

was the *average* mass of the atoms in ordinary hydrogen gas, and if the latter contained a heavy isotope of mass 2, present in the proportion of 1 part in 5000, the discrepancy would be explained. (In the mass spectrograph only the light hydrogen atom contributes to the measurement.) The suggestion was confirmed soon after by Urey, Brickwedde and Murphy (1932), who found faint lines in the spectrum of hydrogen. The wavelengths of these lines were in exact agreement with those calculated for the Balmer series of hydrogen with mass number 2.

(c) The constancy of the speed of light in empty space for all observers in uniform relative motion was first suggested by the experiments of Michelson and Morley between 1881 and 1887. They obtained interference fringes for light propagating in two directions at right angles and found no significant difference for the speed in the two directions, whatever time of the day or year the measurements were made. These and similar measurements led Einstein to the theory of special relativity, one of the great discoveries of physics. Even on pre-relativity theory the difference in the two speeds was expected to be small, and very precise measurements were necessary to show that the difference, if any, was very much less than the expected value. (For a clear account of the Michelson–Morley experiment, see Hecht and Zajac 1974.)

The original measurements of the speed of light c were done with optical waves for which the wavelength is about 500 nm. When measurements were made during and after the Second World War with microwaves - wavelength of the order of 10 mm - the value of c was found to be about 17 km s^{-1} higher than the optical value, despite the fact that the quoted standard errors in the two values were about 1 km s^{-1}. The difference is only 1 part in 20 000; nevertheless, had it been genuine the consequences for our present theories of electromagnetism would have been grave.

Again, precise measurements were necessary to resolve the question. Repetition of the optical measurements did not confirm the previous values but gave results in agreement with the microwave values. The results of three representative experiments are given in Table 8.3. In each of the experiments the values of the frequency f and the wavelength λ of a monochromatic source were measured independently, and the value of c obtained from the relation

$$c = f\lambda.$$

The use of lasers (second and third experiments) gives greatly improved precision. The experiment by Bay *et al.* was the first in which the very

Table 8.3. *Results of three experiments to measure the speed of light c*

Year	Experimenter	Range	λ	$c/\mathrm{m\,s^{-1}}$
1958	Froome	microwave	4·2 mm	299 792 500 ± 100
1972	Bay *et al.*	optical	630 nm	299 792 462 ± 18
1973	Evenson *et al.*	infra-red	3.4 μm	299 792 457·4 ± 1·1

high frequency of an optical source (of the order of 10^{14} Hz) was measured directly, instead of being deduced from a measurement of λ and the value of c. The procedure is to relate the optical frequency to a known microwave frequency by the generation of harmonics and the measurement of beat frequencies. A very readable account of the method has been given by Baird (1983).

The best evidence that the value of c is independent of wavelength comes from astronomical data. Radiation from the pulsar in the Crab Nebula has been observed in the radio, optical, and X-ray ranges of the electromagnetic spectrum. When corrections are made for the effects of free electrons in interstellar space, no significant difference has been observed between the times of arrival of the pulses in different ranges of the spectrum. The results put an upper limit of about 1 part in 10^{13} for a variation in the value of c for this range of wavelengths (Bay and White 1972).

Exercises

Make a careful study of the following experiments:

8.1 New experimental test of Coulomb's law (Williams, Faller and Hill 1971).

8.2 Verification of the equivalence of inertial and gravitational mass (Braginsky and Panov 1972, see also Dicke 1961.)

8.3 Redetermination of Newtonian gravitational constant G (Luther and Towler 1982).

8.4 Apparent weight of photons (Pound and Rebka 1960).

8.5 Determination of the Avogadro constant (Deslattes *et al.* 1974).

8.6 Measurement of an optical frequency and the speed of light (Bay *et al.* 1972).

9

Common sense in experiments

In the present chapter we are going to consider some common-sense aspects of doing experiments. They apply to all experiments, from the most elementary and simple to the most advanced and elaborate.

9.1 Preliminary experiment

In a real experiment, as opposed to an exercise, one nearly always does a trial experiment first. This serves several purposes.

(a) The experimenter 'learns' the experiment. Every experiment has its own techniques and routines, and the experimenter needs some training or practice in them. It is nearly always true that the first few measurements in an experiment are not as reliable or useful as the later ones, and it is usually more economical in time to have an initial period for finding out the best way of making the measurements and recording the results.

(b) The various pieces of apparatus are checked to see that they are working properly.

(c) A suitable range for each variable in the experiment is found.

(d) The errors in the different quantities can be estimated. As we have seen, this influences the strategy of the experiment proper, in the sense that more attention is given to those quantities whose errors give the major contributions to the final error.

Points (c) and (d) really add up to saying that any serious experiment must be planned, and that a few trial measurements provide a better basis for a plan than a lot of theory. Of course the plan must be flexible and is usually modified as the experiment goes along. But even the most rudimentary plan is preferable to making one measurement after another just as one thinks of them.

In an exercise experiment the scope for a preliminary experiment is somewhat limited, and you probably will not have time to do the whole experiment first, even roughly. Nevertheless, except in the very simplest experiment, *some* preliminary measurements should always be made and some sort of plan drawn up. This includes deciding what quantities are to be measured and roughly how long is to be spent on each one.

A related point may be made with regard to a piece of apparatus. Make sure that you know how it is to be operated, in the crude sense of knowing what controls what, before you start making systematic measurements. If you are confronted with a spectrometer, for example, make sure before making *any* measurements, that you know how to rotate the prism table and how to rotate the telescope, which knob must be tightened to allow a micrometer screw adjustment to be effective, which vernier scale corresponds to which motion, and so on. If there is a laboratory manual giving information about the apparatus, or a leaflet put out by the manufacturer, read it *first*.

You may think this is all very obvious, and so it is. But it is surprising how many people are lacking in this elementary 'horse-sense' when it comes to experimental work. Sophisticated ways of treating data and avoiding subtle errors are all very well, but they are no substitute for common sense.

9.2 Checking the obvious

If the apparatus is meant to be mechanically firm, and most apparatus is, see that it is not wobbling about. Remember that three points define a plane, provided they are not in a straight line. So three is the best number of legs for a piece of apparatus, and the closer their positions are to an equilateral triangle the better. With more than three the apparatus will wobble when placed on a plane surface, unless the contact points have been made to lie in a plane.

If the base of the apparatus is meant to be level, and mostly it is, *look* and check that it is approximately so. You can always use a spirit-level afterwards if it must be accurately level.

In optical experiments, make sure that all reflecting and refracting surfaces look clean. A quick breath and wipe will often do wonders with cheap apparatus. But do *not* wipe expensive lenses with a cloth or handkerchief. They are made of soft glass, and in addition are often coated with a very thin film - about 100 nm thick - of a mineral salt to reduce reflection at the surface. Such lenses are easily scratched. They must never be touched with fingers and should be covered when not in use. They will seldom need more than a dusting with a soft camel-hair brush, or in extreme cases a careful wipe with special lens tissue.

See that optical components that are meant to be aligned, look as though they are, and that lenses are turned with their faces roughly at right angles to the direction of the beam. It is amazing how often one sees a student struggling with an optical system with some vital lens, covered with a film of grease, several millimetres too high or low and rotated 10° or so from the normal.

If you have to solder a connection in an electrical circuit, scrape the wires first, then make as firm a mechanical joint as possible. See that the solder melts in a healing flow running over the whole junction. Finally, when cold, wiggle the individual wires gently to make sure that the solder has adhered to them all and that you do not have a dry joint.

When you use a galvanometer or similar instrument with several ranges of sensitivity, always start with the switch on the least sensitive range.

When you are assembling electrical apparatus operated from the mains supply, always plug into the mains *last*, and if you have to service any of it, do not rely on the mains switch being off, but *pull the mains plug out.*

9.3 Personal errors

When you are making measurements you should regard yourself as a piece of apparatus subject to error like any other. You should try to be aware of your own particular errors. For example, in estimating tenths of a division in a scale reading some people have a tendency to avoid certain numbers. You can easily test yourself on this, which is perhaps not a very serious matter.

What can be serious, however, are what might be called wishful errors. Anyone is liable to make an occasional mistake in reading an instrument or in arithmetic. But suppose a series of measurements is producing results which you think are too high. You may well find yourself making more mistakes than usual, and the likelihood is that the majority will be in the direction of decreasing the result. Of course, if you do not know beforehand what to expect, you avoid the danger. This is not always possible, but can sometimes be achieved by a change of procedure or by varying the order in which measurements are made.

A related situation is one where measurements are repeated in rapid succession, and you are influenced by the previous values, so that if the first measurement is a misreading, you tend to repeat the mistake. Alternatively, even if there is no misreading, the results are not genuinely independent, and their spread is spuriously small – see p. 120.

In general you will make fewer mistakes if you are comfortable - physically and mentally. It is worth spending a little time, particularly in a lengthy set of measurements, to ensure this.

(a) Apparatus that needs to be adjusted and controls that are operated often should be conveniently placed.

(b) The same is true for instruments that are read often. In general it is more comfortable to read a vertical than a horizontal dial, and still more comfortable if the dial slopes backwards at a small angle.

(c) The general lighting should be good. (Of course in an optical experiment it is worth going to some trouble to *exclude* stray light.)

(d) Ventilation should be adequate. It is quite important that the air in the laboratory should be as fresh as possible - and not too warm.

(e) Finally, there should be somewhere convenient for your notebook, preferably away from sources of water and heat.

9.4 Repetition of measurements

Measurements of a single quantity should be repeated at least once. Such repetition

(a) serves to avoid mistakes in reading instruments and recording the numbers,

(b) provides a means of estimating the error.

But if we are making a series of measurements of a pair of quantities x, y from which we wish to determine say the slope m of the best line, there is no need to measure y several times for each value of x. Once we have made two pairs of measurements, i.e. obtained two points on the line, we have one value for m. We need further values for m, but it is better to obtain them by making more measurements at different values of x, rather than by repeating the measurements at the same values of x.

It is sometimes the case that we measure a set of x, y values and that the function $y(x)$ is not at all like a straight line. Suppose for example that y is the amplitude of oscillation of a simple harmonic system due to an external oscillating force of frequency x. A typical set of x, y values in the neighbourhood of resonance is shown in Fig. 9.1. In this case the points must be sufficiently close together to define the 'best' curve fairly well.* But, as in the straight-line case, there is no need to repeat the

* How close the points have to be to do this depends on the shape of the function and the precision of the measurements. Beginners are so used to measuring straight-line relations, where relatively few points are needed, that when they come to investigate relations such as that shown in Fig. 9.1, they usually do not have enough points.

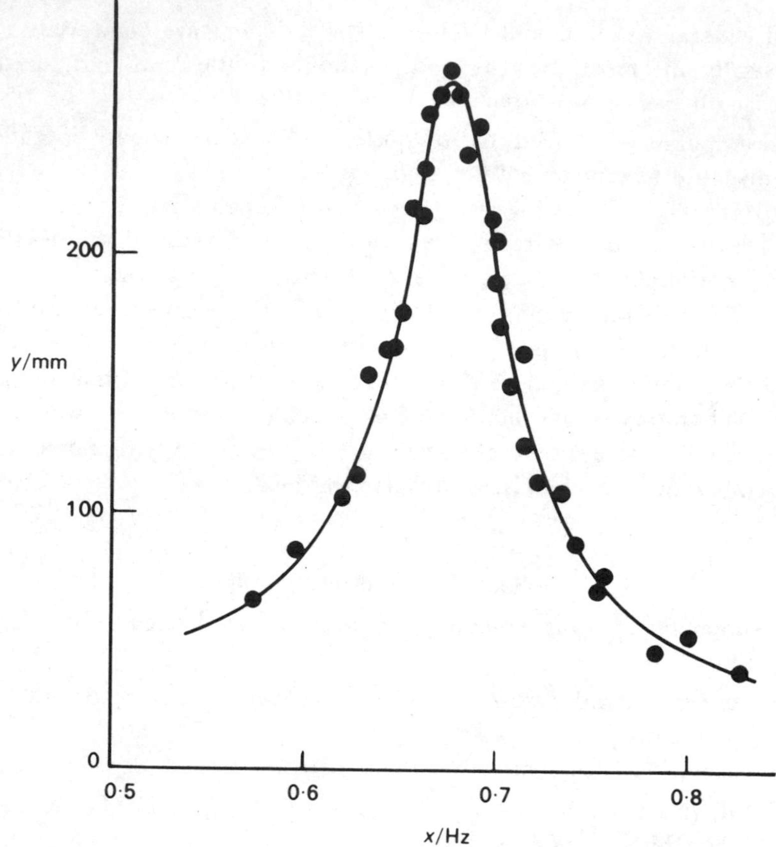

Fig. 9.1. Amplitude of oscillation of a simple harmonic system versus frequency of external force.

measurements. The scatter of the points about the 'best' curve is a measure of their error, though it might be wise to repeat the measurement of y a few times for one or two values of x as a check.

There is one aspect to the repetition of measurements which may be illustrated by the following not-so-hypothetical occurrence. A student in a practical class comes to a demonstrator with a dilemma. He is measuring the angle of a prism with a spectrometer and has obtained the results

$$56^\circ\,30' \quad \text{and} \quad 60^\circ\,12'.$$

He estimates that his accuracy is about $5'$ and, having checked the arithmetic, deduces that one of the results is wrong. (Stop at this point and consider what you would do.)

He asks the demonstrator which one he should take as correct. He is of course being ridiculous. The object of an experiment is to find out something. The student has so far found out nothing, except that one of his answers is wrong. (Quite possibly both are wrong.) In this situation *you must make more measurements.* In fact you must go on making measurements until the results start to make sense. You cannot do anything, not even the simplest manipulation, with the numbers until they *do* make sense.

If the next measurement gave 56° 34′, you would start to think that the second was probably wrong. You should measure the angle once more and, if the result was 56° 35′, you would be fairly certain that this was so. The results are beginning to be sensible. You may still wonder how you obtained the value 60° 12′. Probably you will never find out. Perhaps the prism was inadvertently moved during the measurements, or, more likely, a telescope setting was misread or incorrectly recorded. It is annoying when a wrong result turns up and you cannot account for it, but it does happen, and, provided it is rare, you need not worry. But you should worry if (a) you arbitrarily decided that the prism was a 60° one and therefore the second result was correct, or (b) you decided to take the mean.

9.5 Working out results

We say something about this in chapter 12, but here we only want to make the general point that in any experiment lasting more than a day or two, you should always try to work out the results as you go along.

It is very bad practice to make lots and lots of measurements and to work them out at the end of an experiment. First of all, it is much better to do the calculations while everything is fresh in your mind. Secondly, it is not uncommon to find when you work out a set of results that something is wrong and a change needs to be made to the apparatus. You will be much crosser if this happens after a month than after a day. Quite apart from that, it is often the case that one set of results determines what you do next.

The most foolish thing you can do in this respect is to dismantle elaborate apparatus before working out the results - it has been known.

9.6 Design of apparatus

The principles and techniques of the design of apparatus are discussed in some of the books listed on p. 205. We say very little about the subject here, beyond offering a few words of general advice.

(a) Make things as simple as possible.

(b) If the apparatus is being constructed by a mechanic or instrument maker in the laboratory, discuss the detailed object of the apparatus with him before making definite plans or drawings. His experience may enable him to suggest improvements, or he may see ways of simplifying the apparatus and hence making it easier to construct, without detracting from its performance.

(c) Quote tolerances for the dimensions, and do not make them less than are actually necessary. Obviously, the closer the tolerances to which the constructor is asked to work, the more difficult is his task.

(d) Do some construction yourself. This will give you some idea of what can and cannot be done. An ounce of practice...

PART 3

Record and calculations

10

Record of the experiment

10.1 Introduction

In any experiment it is essential to keep a running record of everything that is done.

The record should be clear - and economical. On the one hand, you do not want to have to spend time subsequently searching pages of numbers without headings to find a particular set of results, or puzzling out from some meagre clues just what the conditions were when you made a certain set of measurements. On the other hand, to produce a record that is so clear that it may be followed with absolute ease by someone else is itself a time-consuming operation and is hardly necessary. You should aim at a record that you yourself will be able to interpret without too much difficulty after an interval of, say, a year.

In this chapter some suggestions for keeping the record are given. The important thing is not that you should regard them as a set of rules to be followed blindly, but rather that you should understand the spirit behind them, which is to produce a record - accurate, complete, and clear - with a minimum of effort.

10.2 Bound notebook versus loose-leaf

Some experimenters prefer a bound notebook; others use loose sheets. The advantage of a single bound book is that one knows where everything is - in the book. There are no loose bits of paper to be mislaid. The main disadvantage is that in an experiment of even moderate complexity one often goes from one part to another, and it is tiresome to have the various parts split into fragments in the record.

The advantage of loose-leaf is its flexibility. All the sheets on a particular topic may be brought together irrespective of what was written in between. The flexibility is useful in another way. In practical work it is convenient to use different kinds of paper - plain, ruled, graph, and data paper. (The last is paper ruled with vertical lines and is very convenient for tabular work.) These can be inserted in any quantity and in any order in a loose-leaf book.

It is best not to be dogmatic about the basic method of keeping the record but to adapt the procedure to the particular experiment. A combination of bound book and loose-leaf may secure the advantages of both. Whatever the system adopted, it is a good idea to have at least *one* bound notebook; it provides a place for odds and ends - occasional ideas, miscellaneous measurements, references to the literature, and so on. It is helpful if the pages of the notebook are numbered, and a detailed list of the contents compiled at the beginning or end of the book.

Experiments done by students as exercises are usually sufficiently short and straightforward to make a combination of bound notebook and loose-leaf sheets unnecessary. Opinion varies as to which is preferable, but the experience of the writer is that the flexibility of loose-leaf gives it the advantage. Different kinds of paper can be used in any order, and the account of a previous experiment may be handed in by the student for critical comment while he carries on with another.

10.3 Recording measurements

All measurements should be recorded *immediately* and *directly*. There is no exception to this rule. Do not perform any mental arithmetic - even the most trivial - on a reading before getting it down on paper. For example, suppose numbers appearing on an ammeter have to be divided by 2 to bring the readings to amperes. Write down the reading given by the instrument markings first. Do not divide by 2 and then write down the result. The reason for this is obvious. If you make a mistake in the mental arithmetic, you are never going to be able to correct it.

In making and recording a measurement, it is a good idea to check what you have written by looking at the instrument again. So

read, write, check.

Note the serial number of any instrument or of any key piece of apparatus, such as a standard resistor, used in the measurements. If the maker has not given it a readily observed serial number, you should give it one of your own. The subsequent identification of a particular piece of apparatus may be important. For example, something may go wrong in the experiment and the inquest may lead you to suspect an erratic instrument. You will want to know which particular one you used.

All written work should be dated.

10.4 Down with copying

An extremely bad habit of many students is to record the observations on scrap paper or in a 'rough' notebook and then to copy them into a 'neat' notebook, throwing away the originals. There are three objections to this:

(a) It is a gross waste of time.
(b) There is a possibility of a mistake in the copying.
(c) It is almost impossible to avoid the temptation of being selective.

The last is the most important and is worth considering further. In most experiments we do not use all the measurements. We often decide that some measurements are not very useful, or were made under the wrong conditions or are simply not relevant. In other words, we are selective. This is quite proper, provided we have objective reasons for the selection. *But it is vital that all the original measurements be retained.* We may subsequently wish to make a different selection. And in any case all the experimental evidence must be available, so that someone else may form an opinion on the validity of our selection, or indeed on any aspect of the original measurements.

An important part of a practical physics course is to train you to keep clear and efficient records, but this training cannot begin until you try to make direct recordings. They will probably look very messy at first and perhaps will be difficult to follow, but do not let that discourage you. You will gradually learn from experience and improve. The result will never look as pretty as the copied-out version, but that is of no consequence. The important thing in a record is clarity, not beauty.

Having said that, let us add that a *certain amount* of copying out may well be useful. It is often a distinct aid to clarity, which in turn is desirable, not only for its own sake, but also because it reduces mistakes in working out the results. It is very often the case that at a certain point in the experiment we want to bring together various results dotted about in different parts of the account. We may want to plot a graph, do some calculations or perhaps just look at the numbers. But, since we are retaining the original readings, this copying may be, in fact should be, highly selective and has nothing to do with the wholesale copying mentioned above.

10.5 Diagrams

'One picture is worth a thousand words' - Chinese proverb.

The importance of diagrams in a record or an account can hardly be exaggerated. Combined with a few words of explanation, a diagram is often the easiest and most effective way of explaining the principle of an experiment, describing the apparatus, and introducing notation. Consider the following alternative descriptions of a piece of apparatus for investigating the motion of a pair of coupled pendulums:

Description 1. A piece of string was fastened to a horizontal rod at two points A and B. Two spheres S_1 and S_2 were suspended by strings, the upper ends of which were attached to the original string by slip knots at the points P_1 and P_2. The lengths AB, AP_1, BP_2, P_1P_2 are denoted by a, y_1, y_2, and x respectively. The distance from P_1 to the centre of S_1 is denoted by l_1, and from P_2 to the centre of S_2 by l_2.

The degree of coupling between the pendulums was varied by varying the distance x. This was done by moving the knots P_1 and P_2 along the string AP_1P_2B, the system being kept symmetric, i.e. $y_1 = y_2$.

Description 2. The apparatus was set up as shown in Fig. 10.1. AP_1P_2B is a continuous length of string.
P_1, P_2 are slip knots.
Coupling varied by varying x by means of slip knots ($y_1 = y_2$ throughout).

Comment on the two descriptions is superfluous.

A diagram should not be an artistic, or even a photographically true,

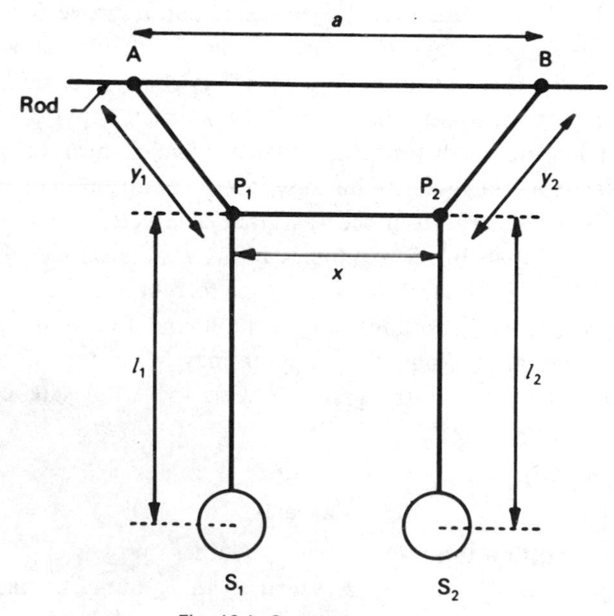

Fig. 10.1. Coupled pendulums.

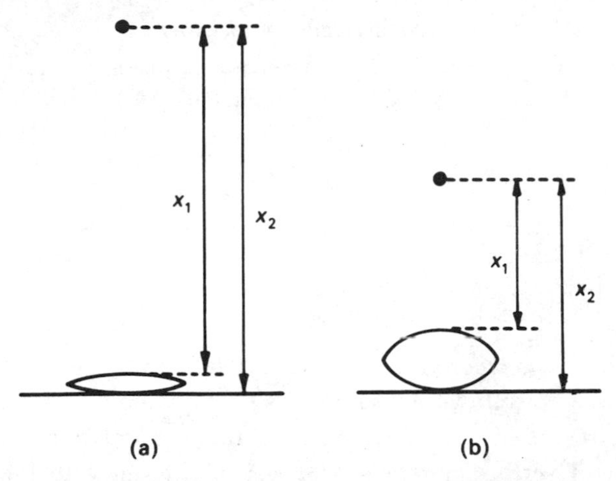

Fig. 10.2. Diagrams for coincident-image method of finding the focal length of a lens.

representation of the apparatus. It should be schematic and as simple as possible, indicating only the features that are relevant to the experiment. Furthermore, although an overall sketch of the apparatus drawn roughly to scale is often helpful, you should not hesitate to distort the scale grossly in another diagram if this serves to make clear some particular point.

Suppose, to take a simple example, we are measuring the focal length of a convex lens by placing it on a plane mirror and observing when an object and its image are coincident. We wish to indicate whether a particular measurement refers to the distance from the object to the top or bottom of the lens. Figure 10.2a is to scale; Fig. 10.2b is not, but is much clearer for the purpose.

A diagram is usually the best way of giving a sign convention. Look at Fig. 10.3 which shows the usual convention for representing a rotation by a vector. Expressing this in words is not only more difficult but also less effective.

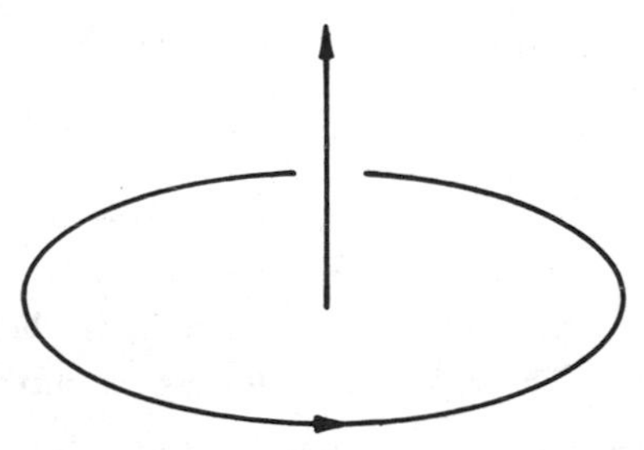

Fig. 10.3. Diagram showing convention for representing a rotation by a vector.

Table 10.1. *Surface tension of water*

T/K	$10^3\,\mathrm{K}/T$	$\gamma/\mathrm{mN\,m^{-1}}$
283	3·53	74·2
293	3·41	72·7
303	3·30	71·2

10.6 Tables

When possible always record measurements in tabular form. This is both compact and easy to follow. Measurements of the same quantity should preferably be recorded vertically, because the eye can more readily compare a set of vertical numbers. Head each column with the name and/or symbol of the quantity, followed by the units.

For convenience, the power of 10 in the unit should be chosen so that the numbers recorded are roughly in the range 0·1 to 1000. There are several conventions for expressing units in a table. The one recommended by the Symbols Committee of the Royal Society* – and used in this book – is that the expression at the head of a column should be a dimensionless quantity. Consider Table 10.1, which gives the values of γ, the surface tension of water, at various values of the temperature T. (The values are taken from Kaye and Laby 1986, p. 43.) We have included also the values of $1/T$ as a further illustration. The quantities T/K, $10^3\,\mathrm{K}/T$, and $\gamma/\mathrm{mN\,m^{-1}}$ are dimensionless, and are therefore suitable headings for the columns. The first line of numbers is interpreted thus.

$$\text{At} \qquad T/\mathrm{K} = 283, \qquad \text{i.e. } T = 283\ \mathrm{K}, \tag{10.1}$$

the value of $1/T$ is given by

$$10^3\,\mathrm{K}/T = 3{\cdot}53, \qquad \text{i.e., } 1/T = 3{\cdot}53 \times 10^{-3}\,\mathrm{K}^{-1}, \tag{10.2}$$

and at this temperature the value of γ is given by

$$\gamma/\mathrm{mN\,m^{-1}} = 74{\cdot}2, \qquad \text{i.e. } \gamma = 74{\cdot}2\ \mathrm{mN\,m^{-1}}. \tag{10.3}$$

Once the unit has been specified at the top of a column it is not necessary to repeat it after each measurement. In general all unnecessary repetition should be avoided. It wastes time, it wastes energy, and it clutters up the record. The fewer inessentials we have, the easier it is to follow the essentials.

* *Quantities, Units, and Symbols*, The Royal Society, 1975.

10.7 Aids to clarity

Diagrams and tables are two aids to clarity. But any such aid is to be welcomed. Groups of measurements of different quantities should be well separated and each should have a title. If a set of measurements leads to one value, say the mean, it is helpful if this one number is not only labelled but also underlined, or made to stand out in some way.

In general you should not be too economical with paper in keeping the record. You will often find yourself starting to record measurements without giving them a title or specifying the units. The habit of leaving a few lines of space at the top permits these items to be added tidily later on. Omitting titles to start with is not necessarily a sign of impatience, but is in fact a sensible practice. You will usually be able to add much more useful titles - often of an extended nature - after several sets of measurements have been made.

A definite hindrance to clarity is the habit of overwriting. Is 37, 27 or 37? Do not leave the reader - or yourself at a later date - to puzzle it out. Just cross out and rewrite ~~27~~ 37.

10.8 Some common faults – ambiguity and vagueness

Example 1. A student is told to measure the viscosity of water at 20 °C and to compare the value obtained with that given in some table of physical constants. The following appears in his notebook:

experimental value	$1{\cdot}005 \times 10^{-3}$ N s m^{-2}
correct value	$1{\cdot}002 \times 10^{-3}$ N s m^{-2}

Which is his value and which the one given in the tables? If we know him to be a modest person we might guess that what he calls the 'experimental value' is his own, and what he calls the 'correct value' is the one in the tables. If he is conceited it might be the other way round. But of course we ought not to have to guess, on the basis of his personality or otherwise. He should have written something like

this experiment	$1{\cdot}005 \times 10^{-3}$ N s m^{-2}
Kaye & Laby (15th ed. p. 36)	$1{\cdot}002 \times 10^{-3}$ N s m^{-2}

Other adjectives such as 'actual', 'official', 'measured', 'true' are equally vague in this context and should be avoided.

Note that a detailed reference to the tables has been given. We might want to check the value or look at the table again.

Example 2. Consider a notebook entry such as

Ammeter A 14 zero error $-0{\cdot}03$ A

Does this mean that when no current passed the instrument read $-0{\cdot}03$ A, and therefore 0·03 should be *added* to all the instrument readings to get the correct value, or does it mean that 0·03 should be *subtracted*? Again we are left to guess.

In accordance with the rule that measurements should be written down directly, without the intervention of any mental arithmetic, what the experimenter should do is to read the instrument when no current is passing and write down the value. So the entry should be something like

Ammeter A 14
−0·03 A ← reading when no current passed

Example 3. The following is a statement of a kind often found in a student's notebook:

'The timebase of the oscilloscope was checked with a timer-counter and found to be accurate within experimental error.'

This is objectionable for its vagueness on two essential points. Firstly, an oscilloscope timebase has a number of ranges and the statement does not indicate which one or ones were checked. Secondly, the *evidence* for the statement has not been given, and without that evidence we have no means of knowing whether the statement is justified. What should have appeared is something like this:

Oscilloscope SC29 – calibration of timebase range 0·1 ms cm^{-1}
sine-wave applied to Y plates f = 10·018 kHz (timer-counter)

oscilloscope	cycle	x/cm	5 cycles
	0	0·95	5·00
	1	1·96	4·92
	5	5·95	4·96 ± 0·04 cm
	6	6·88	

$$\text{sweep speed} = \frac{5}{10\cdot018 \times 4\cdot96} = 0\cdot1006 \pm 0\cdot0008 \text{ ms cm}^{-1}$$

Conclusion timebase correct within experimental error for this range

The above examples illustrate the following points:

(a) What is stated must be unambiguous. You should consciously ask yourself whether any interpretation other than the correct one is possible. Quite often the simplest way to resolve possible ambiguity is to give a single numerical example.

(b) If a conclusion is based on numerical evidence, and nearly all conclusions in physics experiments are or should be so based, then *the numbers must be given explicitly.*

11

Graphs

11.1 The use of graphs

In experimental physics, graphs have three main uses. The first is to determine the value of some quantity, usually the slope or the intercept of a straight line representing the relation between two variables. Although this use of graphs is often stressed in elementary teaching of practical physics, it is in fact a comparatively minor one. Whether we obtain the value of the slope of a straight line by the method of least squares or by taking the points in pairs (see section 4.2), we are of course not using the graph as such, but the original numbers. The only time we actually use the graph to determine the slope is when we judge or guess the best line through the points by eye. This is a crude method - though not to be despised on that account - and should only be used as a check on the result of a more sophisticated method, or when the value of the slope is not an important quantity in the final result.

The second use of graphs is by far the most important. They serve as *visual aids.* Suppose, for example, the rate of flow of water through a tube is measured as a function of the pressure gradient, with the object of determining when the flow ceases to be streamlined and becomes turbulent. A set of readings is listed in Table 11.1. (They are taken from Reynolds' original paper on turbulent flow - Reynolds 1883.) As long as the flow is streamlined, the rate of flow is proportional to the pressure difference. It is difficult to tell by inspecting the numbers in the table where the proportionality relationship breaks down. However, if the numbers are plotted on a graph (Fig. 11.1), the point of breakdown is apparent at once.

Another example of this visual use is the comparison of experimental results with a theoretical curve, when both are plotted on the same graph. In general a graphical display of the measurements is invaluable for showing what is going on in an experiment.

The third use of graphs in experimental work is to give an empirical relation between two quantities. For example, a thermometer may have

Table 11.1. *Flow of water through a tube*

Pressure gradient/Pa m^{-1}	Average velocity/mm s^{-1}
7·8	35
15·6	65
23·4	78
31·3	126
39·0	142
46·9	171
54·7	194
62·6	226
78·3	245
86·0	258
87·6	258
93·9	271
101·6	277
109·6	284
118·0	290

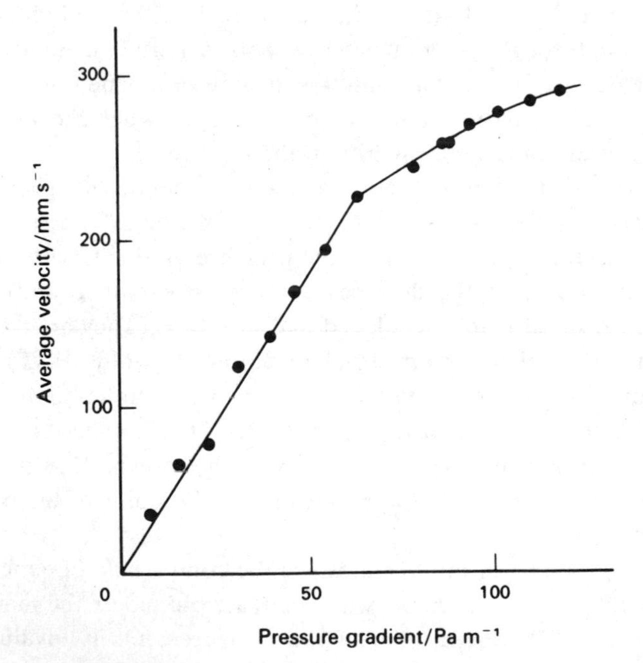

Fig. 11.1. Average velocity of water through a tube versus pressure gradient – plot of values in Table 11.1.

been calibrated against some standard, and the error determined as a function of the thermometer reading (Fig. 11.2a). We draw a smooth or average curve through the measured errors and use it (Fig. 11.2b) to correct the thermometer readings. We could have done the same thing by compiling a correction table. In general a table is more convenient to use than a graph, but may be more trouble to compile.

It is a well-established convention for all graphs in physics that the independent variable, i.e. the quantity whose value is chosen each time by the experimenter, is plotted along the horizontal axis, and the dependent variable, i.e. the quantity whose value is then determined, plotted along the vertical axis. Or, more briefly, plot *cause* along the horizontal and *effect* along the vertical axis.

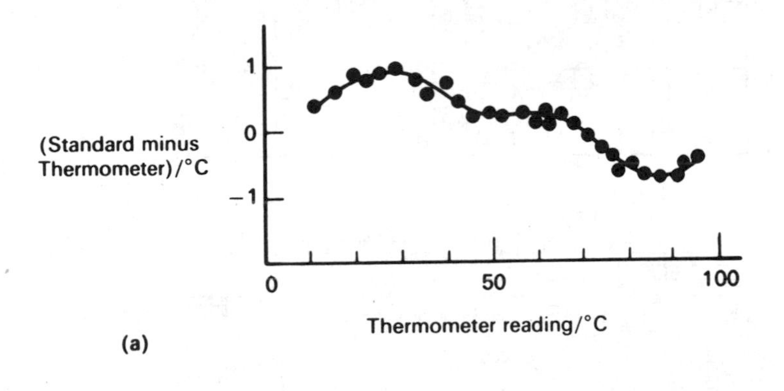

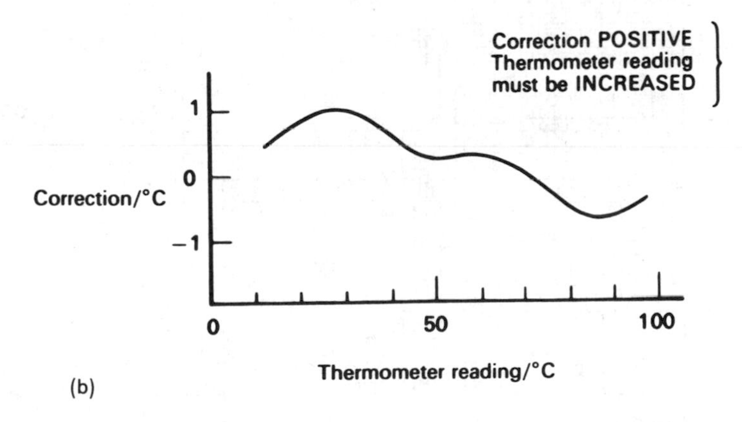

Fig. 11.2. (a) Calibration measurements for a thermometer and (b) correction curve.

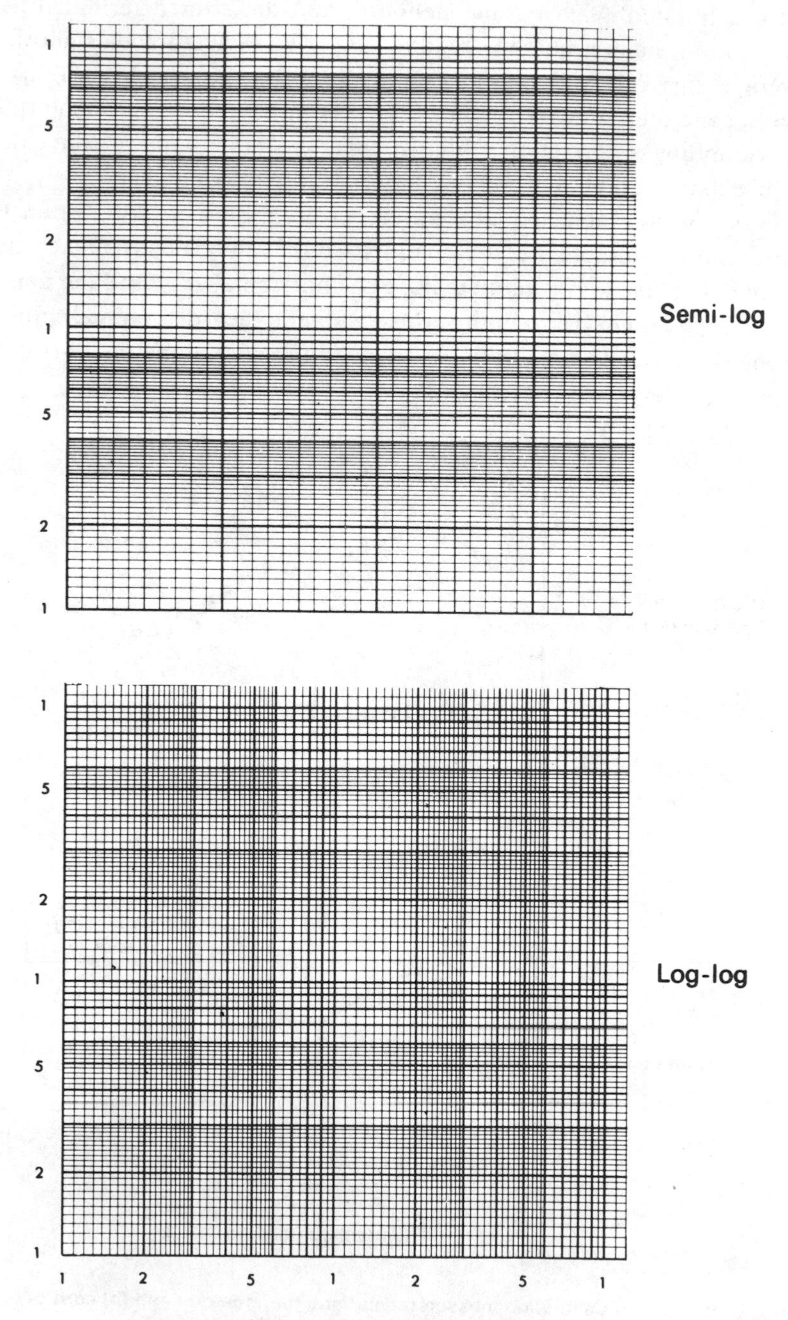

Fig. 11.3. Graph paper with logarithmic rulings.

11.2 Choice of ruling

Graph paper can be obtained in a variety of rulings for specialized purposes, but the two most commonly used in physics are ordinary linear and logarithmic rulings. The latter may be further subdivided according to whether only one axis is ruled logarithmically (semi-log) or both axes (log–log) - see Fig. 11.3. Semi-log paper is useful when there is a logarithmic or exponential relation between the two variables. Log–log paper is useful when the relation is of the form

$$y \propto x^p,$$

and the value of p is not known.

11.3 Scale

Suppose we are using graph paper ruled in centimetres and millimetres. Our choice of scale should be governed by the following considerations:

(a) The experimental points should not all be cramped together. It is rather difficult to get much out of Fig. 11.4a. So we choose a scale to make the points cover the sheet to a reasonable degree as in Fig. 11.4b. However, in trying to do this we should bear in mind two further points.

(b) The scale should be simple. The simplest is 1 cm on the graph paper representing 1 unit (or 10, 100, 0·1, etc.) of the measured quantity. The next simplest is 1 cm representing 2 or 5 units. Any other scale should be avoided simply because it is tedious to do the mental arithmetic each time a point is inserted or read off.

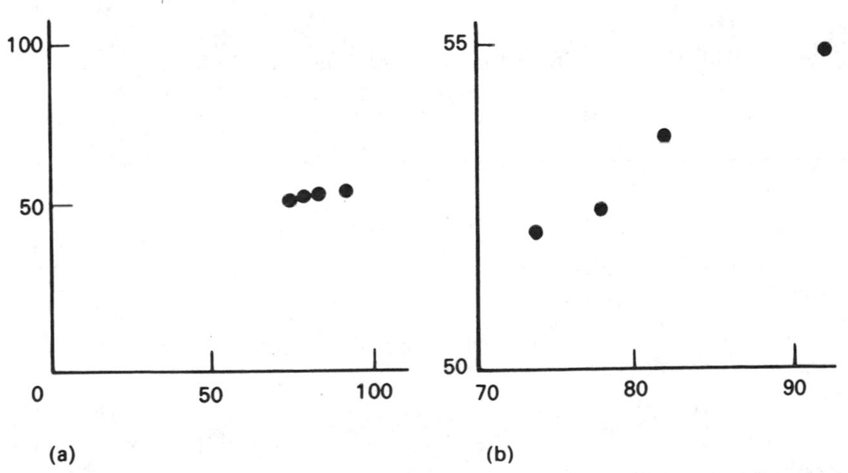

Fig. 11.4. (a) is not a very useful graph. The same results are plotted on an expanded scale in (b).

(c) We sometimes have to choose the scale for theoretical reasons. Thus if we are investigating whether the results in Fig. 11.4 satisfy the relation $y = mx$, we must include the origin in a plot of y against x, and Fig. 11.4b is wrong. (This does not mean that we need revert to Fig. 11.4a - see section 11.7.)

11.4 Units

It is usually convenient to choose the power of 10 in the unit on the same basis as for tables (see section 10.6). The marks on the graph can then be labelled 1, 2, 3, ... or 10, 20, 30, ... rather than 10 000, 20 000, etc., or 0·0001, 0·0002, etc.

The axes should always be labelled with the name or symbol or both of the varying quantity. The unit should be given with the same convention as in tables - see p. 144. Some examples are given in Fig. 11.5.

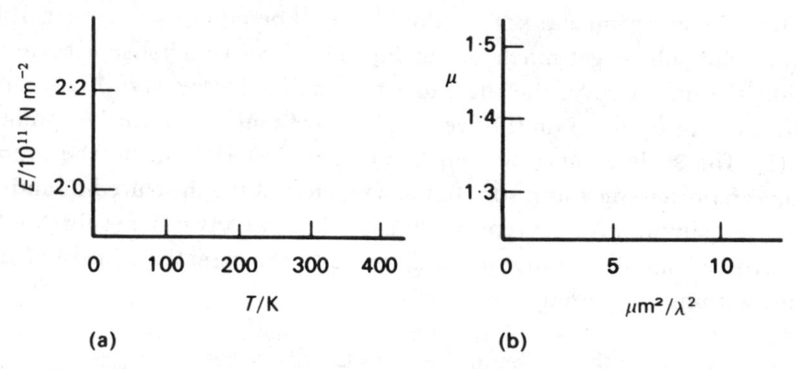

Fig. 11.5. Examples of labelling axes and expressing units. (a) Young modulus E versus temperature T. (b) Refractive index μ of a glass versus $1/\lambda^2$, where λ is the wavelength of the light.

11.5 Some hints on drawing graphs

The main purpose of a graph is to give a visual impression of results, and it should therefore do this as clearly as possible. We give here a few general hints for drawing graphs. They should be interpreted and modified according to the particular case.

(a) If a theoretical curve is drawn on the graph for comparison with the experimental results, the calculated theoretical points through which the curve is drawn are chosen arbitrarily. They should therefore be inserted faintly, preferably in pencil, so that they can be rubbed out.

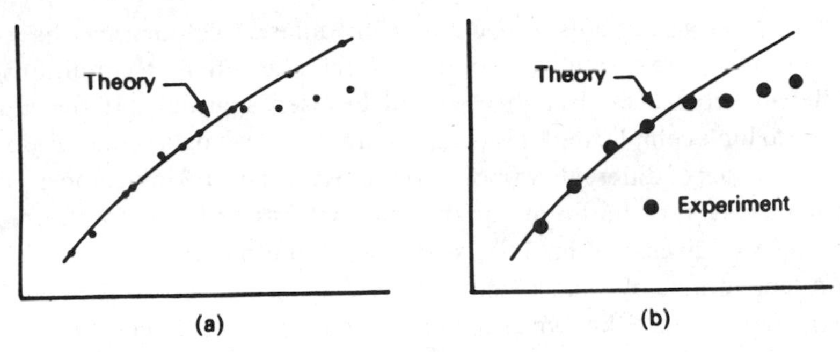

Fig. 11.6. (a) is a poor graph – the experimental points are faint and indistinguishable from the points calculated for drawing the theoretical curve. In (b) the calculated points have been erased, and the experimental points are prominent.

On the other hand, each experimental point should be represented by a bold mark – not a tiny point – so that it stands out clearly – Fig. 11.6.

(b) It sometimes helps if a 'best' smooth curve is drawn through the experimental points. Note the word *smooth.* Beginners sometimes join up experimental points as in Fig. 11.7a. But this implies that the relation between the two variables has the jagged shape shown, which except in special circumstances is highly unlikely. We expect the relation to be represented by something like the curve shown in Fig. 11.7b.

If a theoretical curve is also drawn on the graph, it is usually better to omit the curve through the experimental points. The curve may imply more than the actual results warrant and distracts from the direct comparison of experiment and theory.

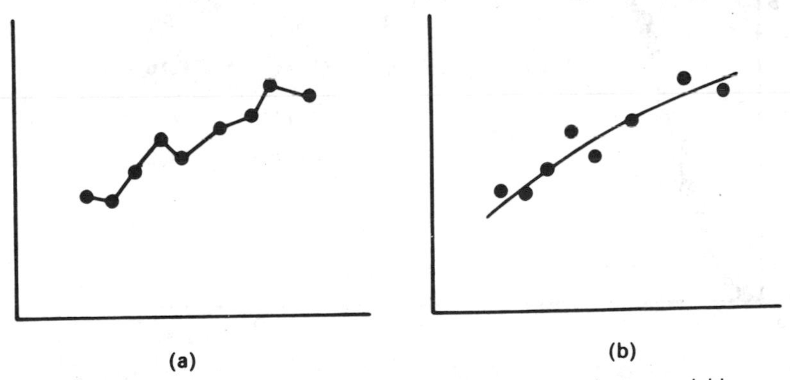

Fig. 11.7. (a) is wrong – it implies that the relation between the two variables has the jagged shape shown, which is most unlikely. From the experimental values we expect the relation to be something like the curve in (b).

(c) Different symbols, e.g. ●, ○, ×, or different colours may be used to distinguish experimental points that refer to different conditions or different substances. But they should be used sparingly. If the graph starts to look complicated it is better to draw each set on a separate graph.

The device of different symbols is perhaps most useful in demonstrating that varying the conditions or the material has little or no effect. An example is given in Fig. 11.8, where C_{mV}, the molar heat capacity at constant volume of a substance, is plotted against T/θ. T is the absolute temperature, and θ, known as the Debye temperature, is a constant which depends on the substance. According to the Debye theory of specific heats, the relation between C_{mV} and T/θ is the same for all solids. Some results for lead ($\theta = 88$ K), silver ($\theta = 215$ K), copper ($\theta = 315$ K) and diamond ($\theta = 1860$ K) are given in the figure, together with the form of the relation predicted by Debye. It can be seen that for these substances the experimental results are in good agreement with the theory.

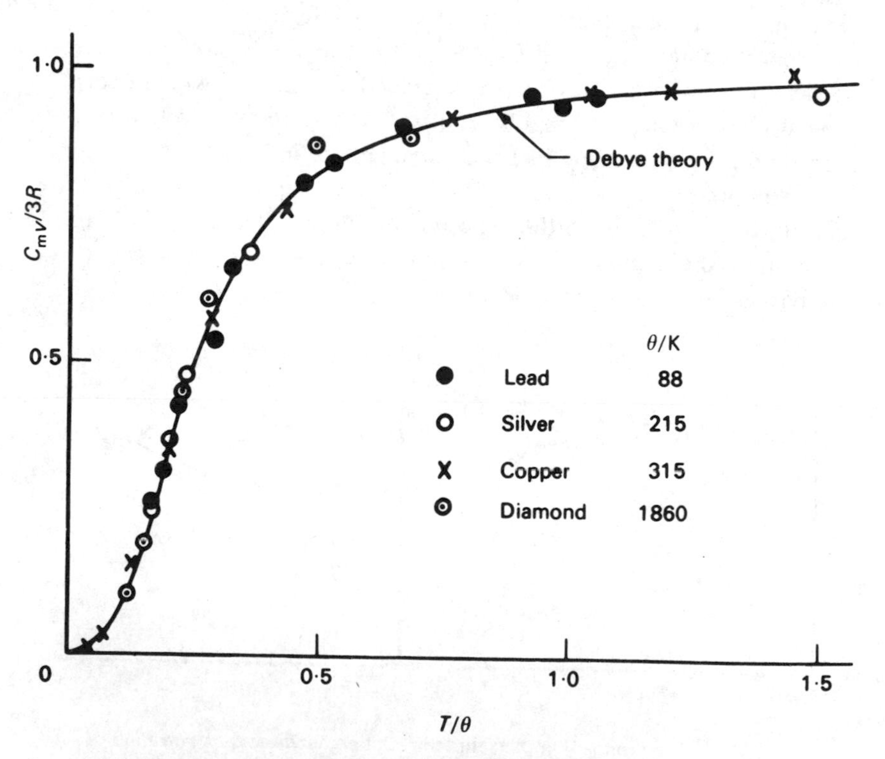

Fig. 11.8. Molar heat capacity, C_{mV}, in units of $3R$, versus T/θ for lead, silver, copper, and diamond.

Notice that the quantity plotted along the y-axis is $C_{mV}/3R$, where R is the gas constant. It is a common procedure in physics to express a physical quantity in dimensionless form by means of some natural unit. In the present case the unit, $3R$, is the value of C_{mV} predicted by classical theory and also by the Debye theory in the high-temperature limit ($T \gg \theta$).

(d) It is a good idea to mark out the scale along the axes and to insert the experimental points in pencil in the first instance. You will sometimes change your mind about the scale and occasionally put a point in the wrong place initially. When you are satisfied with the scale and position of the points, it is easy to ink everything in and draw bold experimental points. The practice avoids messy alterations or wasting graph paper through redrawing.

11.6 Indicating errors

The error in an experimental point may be indicated thus

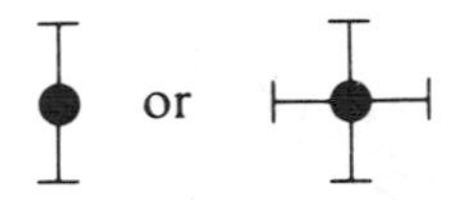

Since inserting the error bars is an additional labour and complicates the graph, it should only be done if the error information is thought to be relevant. Little would be gained for example by adding the error bars to the points in Figs. 9.1 or 11.8.

On the other hand, the significance of deviations from a theoretical curve may depend on the estimated error, and in that case the errors should be indicated. Thus in Fig. 11.9a the deviations would not be considered significant, whereas in Fig. 11.9b they would. We have already encountered this situation in section 8.6, where the spread of a set of results is not consistent with their errors. In that case the theoretical curve is the straight line

$$\text{speed of sound} = \text{constant}.$$

Plotting the experimental results together with the estimated error – Fig. 8.3 – is a useful way of showing the discrepancy.

Another situation in which errors are commonly shown is when they are different for the various experimental points.

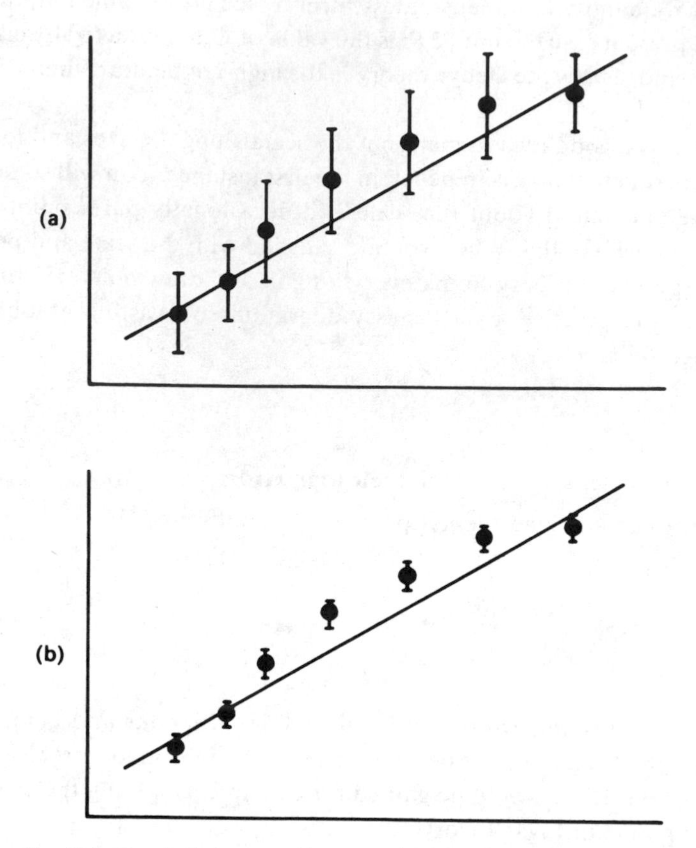

Fig. 11.9. The deviations are the same in the two figures, but in (a) they are probably not significant, whereas in (b) they probably are.

11.7 Sensitivity

Suppose we are doing an experiment to determine whether a certain equation

$$y = x$$

is valid. We obtain pairs of values for y and x and find that the equation is approximately true. If we wish to show the results graphically, we can plot y against x - Fig. 11.10a. However, it is much more sensitive to plot $y - x$ against x, because $y - x$ is small compared to y, and we may use

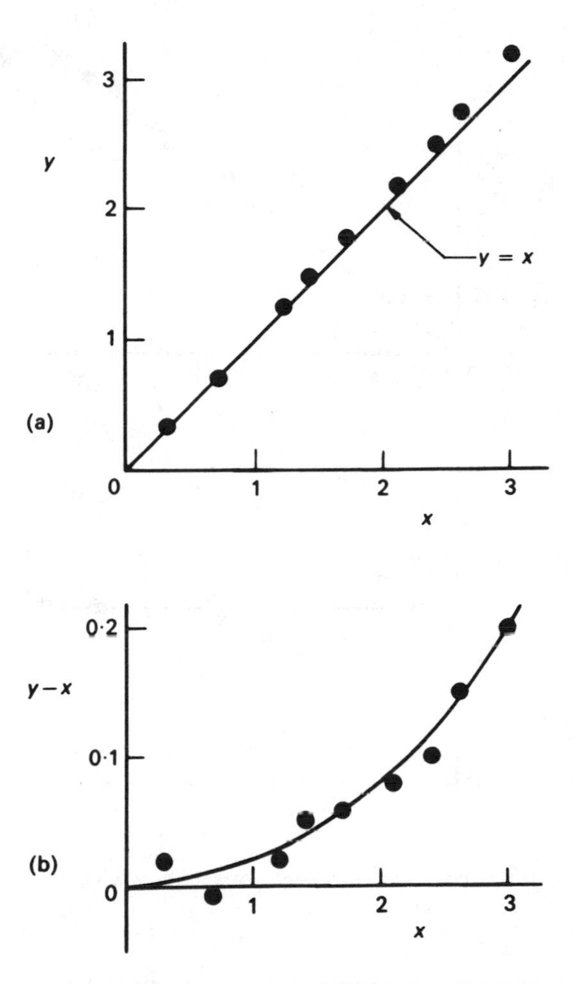

Fig. 11.10. (a) y versus x, and (b) $y-x$ versus x.

a much expanded scale for it - Fig. 11.10b. The departure from the equation $y=x$, slightly indicated in the first figure, is quite evident in the second. Figure 11.2a is an example of this way of plotting results.

A similar method is applicable to the relation

$$y = mx.$$

A direct plot of y against x gives an overall picture of the relation and

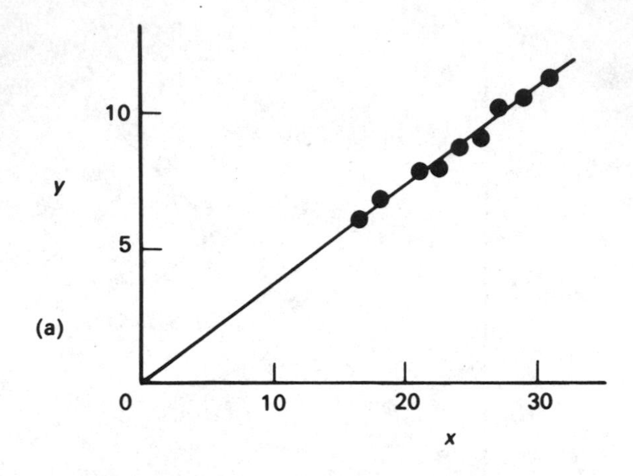

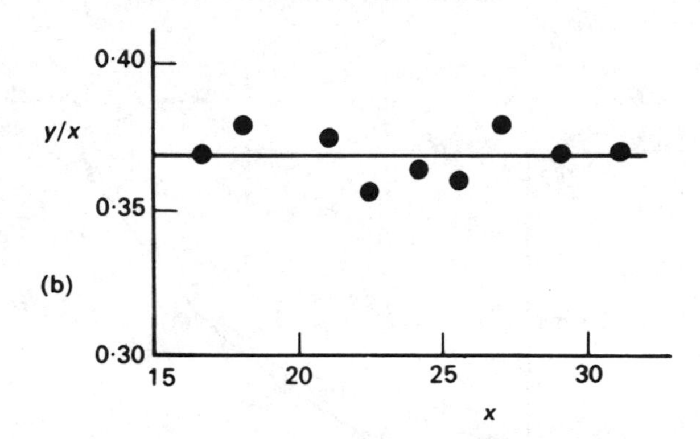

Fig. 11.11. (a) y versus x, and (b) y/x versus x.

may be useful on that account - Fig. 11.11a. But plotting y/x against x is much more sensitive. We do not have to include the origin as in the direct plot, but may have whatever range of values for y/x and x is convenient - Fig. 11.11b.

12

Arithmetic

12.1 Arithmetic is important

The object of an experiment is to obtain a number, and the correct working out of that number is just as important as the taking of the measurements. Many experiments performed by students, containing sensible measurements, are ruined by mistakes in calculating the results.

The following devices are available for calculations:

computer,
calculator,
you.

They are listed in order of decreasing expense and increasing availability. Choose the one appropriate to the job.

12.2 Computers

Computers are outside the scope of this book,* but we make a few general points. Computers range from large main-frame machines, designed to serve many users, to small microcomputers that are now so cheap that they are found in many homes. There are experiments involving the accumulation and interpretation of large amounts of data where large computers are necessary, for example in the processing of radio and optical images in astronomy, or in the determination of complicated biological structures by the analysis of X-ray diffraction patterns. However, for the type of experiments we are concerned with microcomputers are more than adequate. For example, the calculation of the best straight line and its error by the method of least squares is a trivial operation for such a computer.

Apart from their use as calculating devices, computers have an important role in the experiment itself – in the control of apparatus and the

* See p. 205 for some introductory books on the subject.

recording of data. It is now common practice in research experiments to have a minicomputer or a microcomputer dedicated to a particular instrument. A microprocessor is the central processing unit of a small computer, i.e. the part that does the actual arithmetic and logic operations. It may comprise only a single chip, and it can be built directly into an instrument or piece of equipment to give digital readings and control. You will almost certainly use microprocessors and computers in experimental work, both in your training and later. However, it is vital that you know exactly what you want to do in an experiment, how it should be planned, what are the main sources of error, in short all the things this book is about, *before* you start designing the microprocessor or computer part of the experiment.

12.3 Calculators

An electronic calculator, which is essentially a small portable computer, is so convenient, accurate and versatile, that it has become the standard tool for doing arithmetic in experimental work.

All calculators do the four basic operations of addition, subtraction, multiplication and division. It is worth getting what is termed a *scientific calculator*, which also calculates squares, square roots, reciprocals and the common functions - logarithmic, exponential and trigonometric. A number of storage registers, or memories, is also an advantage. Some calculators have built-in programs for calculating the mean and standard deviation of a set of numbers, and the best straight line through a set of points. A calculator that you can program yourself is an advantage for repetitive calculations.

Most calculators produce numbers with up to 8 or 10 digits. This is of course far more than are significant for the vast majority of experiments. You should avoid the elementary blunder of quoting values with a large number of meaningless digits just because they are given by the calculator. Many calculators have the facility that you can fix the number of digits after the decimal point. This is very convenient and should certainly be used if available. It is much easier to appreciate the significance of numbers, and you will make fewer mistakes, if you eliminate most of the insignificant digits.

You should always retain at least one and possibly two digits beyond those that are significant. You should never so truncate a number that you lose information. So if you are not sure at the time how many digits *are* significant, err on the side of caution and include more rather than less.

With the advent of the calculator, mathematical tables have largely fallen out of use, except for specialized functions such as Bessel and Legendre functions, and for functions such as the integral function (p. 176) which are not readily obtained from a calculator.

12.4 Ways of reducing arithmetical mistakes

It might be thought that with calculators and computers there is no need to worry about arithmetical mistakes. Experience suggests otherwise. It is true that calculators and computers are very reliable. They will nearly always give the right answer based on what you tell them. The trouble is that what you tell them may be wrong. You may enter a wrong number, or press a wrong function key, or make a mistake in program logic. Some students appear to have the view that mistakes of this kind are acts of God quite beyond the control of people making them. The answer to that is that though anyone is liable to make such mistakes, first it is possible to reduce their likelihood by sensible procedures, and secondly, there is a remedy, namely, to *check* the calculation. We consider these two points in turn.

(a) Avoid unnecessary calculations. The less calculating you do, the less opportunity you have for going wrong, and the more mental energy you have available for necessary calculations.

Suppose, for example, you are doing a simple experiment to determine the force constant λ of a spring, defined by the equation

$$F = \lambda x, \qquad (12.1)$$

where F is the applied force in newtons and x the resulting extension in metres. You have a set of weights 1, 2, 3, . . . , 6 kg and you measure the extension produced by each one in turn. Do you then convert each weight into newtons by multiplying by 9·81? You would be very foolish to do so. That would involve six multiplications and, moreover, would convert the six simple integers into six messy numbers. You should of course do all the arithmetic to find the best value of λ, keeping F in the integer form. Only at the end should you multiply the value of λ, in units of kilogram-weight per metre, by 9·81 to get the answer in newtons per metre.

The same point applies when a quantity is measured a number of times with an instrument which has been calibrated against a standard. The calibration correction should be made only to the mean of the measurements, and not to each measurement separately.

(b) Be tidy. Calculations should be set out as systematically and tidily as possible. Space the working liberally. Untidy and cramped calculations are a prolific source of mistakes.

Most of the points made in connection with recording measurements apply also to calculations. A tabular arrangement for numbers is often the most convenient and effective. Very often the numbers in one column are the results of manipulations of numbers in one or more previous columns. Every column should have a heading of some kind to indicate the manipulation. These headings are often simplified if the columns are in addition labelled alphabetically. For example, suppose you are calculating a function that occurs in resonance theory

$$y=\frac{1}{Q}\left[(1-x)^2+\frac{x}{Q^2}\right]^{-\frac{1}{2}} \tag{12.2}$$

for the value $Q=22$. A simple set of table headings would be

A	B	C	D	E	F
x	$(1-x)^2$	$\frac{x}{484}$	B+C	$1/\sqrt{\text{D}}$	$y=\frac{\text{E}}{22}$.

12.5 Checking calculations

Checking should be regarded as part of the calculation. The experimenter is here in the same situation as a manufacturer of motor cars. The latter must have a department for inspecting the cars before they are sent out. He regards this as a necessary part of the cost of producing a car. In the same way, part of your time and effort must be devoted to checking the calculations. But it is up to you to get the greatest return for the effort, in other words, to direct it where it is most needed. Some calculations in an experiment are more important than others and should therefore be checked more carefully.

Calculations can be divided into two categories, which may be termed 'self-checking' and 'non-self-checking'. Suppose for example we are measuring two physical quantities which after a little arithmetic give a pair of numbers x_i, y_i. The measurements are plotted in a graph and give approximately a straight line. We would be justified here in not checking the arithmetic very carefully for every x_i and y_i, because if we made a mistake it would very likely stand out on the graph - Fig. 12.1. This is an example of a set of self-checking calculations.

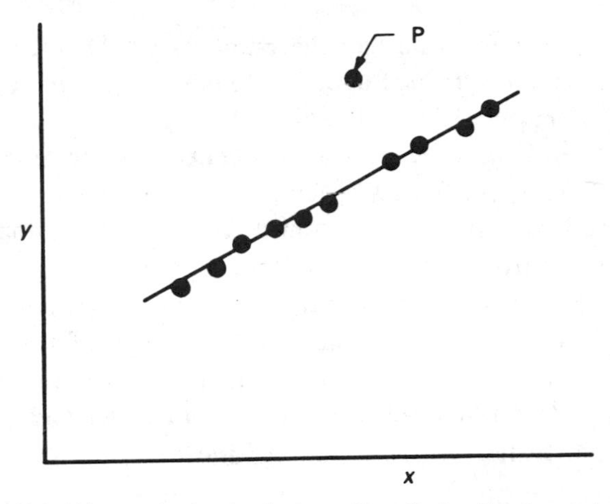

Fig. 12.1. We need not check the arithmetic leading to each point very carefully, because a mistake will probably be obvious – P is almost certainly a mistake.

But suppose we end an experiment with the quantity to be determined given by

$$Z = \frac{14{\cdot}93 \times 9{\cdot}81 \times 873}{6{\cdot}85 \times (0{\cdot}7156)^2 \times \pi^2}. \tag{12.3}$$

The evaluation of this expression is not self-checking. There is nothing to tell us whether the answer is correct, except some method of checking. If a calculator is being used, it is worth doing the calculation twice, particularly if the numbers in (12.3) are not already stored in the calculator memories and have to be keyed in.

You may think it is showing excessive caution to work out the result twice as suggested; but arithmetical mistakes in experiments are a major source of wasted effort. On balance you will save time by careful calculation. Remember that, though in a classroom someone may go through your work and find your mistakes, no one will do so later on. You should get into the habit of making a rough mental estimate - good to about 1 in 3 - of every calculation. Thus for the expression in (12.3) you should say something like

$$\frac{14{\cdot}93}{6{\cdot}85} \approx 2.$$

$$2 \times 9{\cdot}81 \times 873 \approx 20\,000$$

$$(0{\cdot}7156)^2 \times \pi^2 \approx \tfrac{1}{2} \times 10 = 5.$$

So

$$Z \approx 4000. \tag{12.4}$$

So far we have assumed that the numbers in (12.3) are correct, i.e. free from calculation errors. But of course this may not be so, particularly if they are themselves the end results of long calculations. One of the conveniences of a calculator is that it is not necessary to record the results of intermediate steps, and it is tempting not to do so. However, the consequence is that if you make a mistake somewhere along the way it is not possible to trace and correct it. If the repetition of a long calculation gives a different answer you still do not know which one (if either) is correct. It is therefore important to record some of the intermediate values. How many and which ones depend on the complexity of the calculation and your own reliability. You would be well advised to record more rather than less at the beginning, and to reduce the amount with experience.

If a check calculation does not agree with the original, look at the check first, as it was probably done less carefully than the original. A story is told of a new research student in theoretical physics who took the result of an elaborate calculation to his supervisor, a very famous physicist. The supervisor looked at it and said, 'If we take the following special case, your result should reduce to such and such.' He scribbled a couple of lines of calculation on the back of an envelope and said 'You see, it doesn't. You've gone wrong somewhere.' The downcast student took his work away and spent the next month going through it all again. He sought his supervisor once more. 'Well,' said the great man, 'did you find the mistake?' 'Yes,' replied the student, 'it was in your two lines of calculation.'

Look at all numerical values to see if they are reasonable. If you divide instead of multiply by the Planck constant in SI units, you should notice that the answer looks wrong. Of course, to know whether a value is reasonable or not you must have some idea of the orders of magnitude of various physical quantities in the first place - see exercise 12.1.

12.6 Error calculations

We saw in chapter 5 that error calculations should be done to at most two significant digits. So if you are calculating a standard deviation with a calculator, truncate the value accordingly. You should glance at the individual readings to check that the value of σ is reasonable, i.e. that roughly two-thirds of the readings lie within $\pm\sigma$ of the mean. Alternatively use the range method (p. 26) as a check.

For a function of the form

$$Z = \frac{AB\ldots}{CD\ldots}, \tag{12.5}$$

the fractional error in Z must be larger than the largest fractional error in the measured quantities A, B, C, ..., and you should always check that this is so. On the other hand, it is seldom much larger, and you should check this also. For a function of the form

$$Z = A \pm B \pm C \pm \ldots, \tag{12.6}$$

the same remarks apply to absolute errors.

12.7 Checking algebra

(a) Dimensions. The dimensions of an algebraic expression provide a useful check. It is not worth checking the dimensions at every stage of a theoretical argument, but an expression such as

$$l^2 + l,$$

where l is a length, should leap to the eye as being wrong.

In expressions like $\exp x$, $\sin x$, $\cos x$, the dimensions of x must be zero. This is true of any function that can be expressed as a power series in x.

(b) Special cases. You should check that an expression reduces to the correct form in special simple cases - the very famous physicist had the right idea.

(c) Direction of variation. Make sure that an expression gives a variation in the right *direction* as some component quantity varies. Consider, for example, Poiseuille's equation

$$\frac{\mathrm{d}V}{\mathrm{d}t} = \frac{p\pi r^4}{8l\eta}. \tag{12.7}$$

(The quantities are defined on p. 124.) We expect on physical grounds that if p or r is increased, or if l or η is decreased, then $\mathrm{d}V/\mathrm{d}t$ should increase. The form of the equation is in agreement with each of these variations.

(d) Symmetry. Symmetry sometimes provides a useful check on the correctness of a formula. The resistance of the arrangement in Fig. 12.2

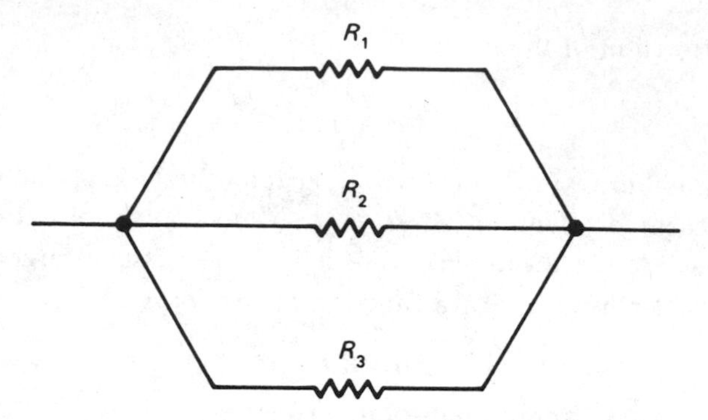

Fig. 12.2. Arrangement of resistors.

is

$$\frac{R_1R_2R_3}{R_1R_2+R_2R_3+R_3R_1}. \tag{12.8}$$

If we exchange any two of the Rs, say R_1 and R_2, we get the same expression, which of course we must, because the arrangement is symmetric in the three resistances. But suppose we had made a mistake in the algebra and had obtained the result

$$\frac{R_1R_2R_3}{R_1(R_2+R_3)}. \tag{12.9}$$

We would know at once that it was wrong, because interchanging R_1 and R_2 gives a different result.

Table 12.1. *Approximations for some functions for $x \ll 1$*

Function	Approximation
$(1+x)^{\frac{1}{2}}$	$1+\frac{1}{2}x$
$\frac{1}{1+x}$	$1-x$
$(1+x)^a$	$1+ax$
$\sin x$	x
$\cos x$	$1-\frac{1}{2}x^2$
$\tan x$	x
$\exp x$	$1+x$
$\ln(1+x)$	x

(e) Series expansion. The first one or two terms of a series expansion of a function often provides a useful approximation. Some common functions are given in Table 12.1.

(f) A useful tip. If you have to do some algebraic manipulation of quantities for which you have numerical values, do all the algebra first and get the final algebraic answer. Substitute the numerical value right at the end. It is much easier to avoid mistakes in this way. Quite apart from that, once you substitute numbers you lose the check on dimensions.

Exercises

12.1 The following exercises are given to test your knowledge of orders of magnitude of various physical quantities. They are in the form of simple problems, because there is not much point in knowing the value of a physical quantity unless you know how to make use of it. You should try the problems in the first instance without looking up any of the values. Try to make an intelligent guess of those you do not know on the basis of any theoretical or practical knowledge you have of related quantities.

When you have gone through all the problems in this way, look up the values in a book like Kaye and Laby before looking at the answers. You will find browsing through Kaye and Laby far more instructive than just checking the answers given here.

(a) A copper rod is 40 mm in diameter and 200 mm long. If one end is at 0 °C, how much heat must be supplied per second at the other end to keep it at 25 °C?

(b) A steel rule is correct at 20 °C. What is its error at 30 °C?

(c) A copper wire of diameter 1 mm is 1 m long. (i) What is its resistance at 0 °C and (ii) by how much does it change between 0 °C and 20 °C?

(d) The cold junction of a chromel–alumel thermocouple is at 0 °C, and the hot junction is at 100 °C. What emf results?

(e) Water flows through a tube 1 mm in diameter and 250 mm long under a pressure difference of 2000 N m^{-2}. What is its average velocity (i) at 20 °C and (ii) at 50 °C?

(f) The cross-section of a rectangular steel bar is 25 mm by 5 mm. If its length is 1 m, what force is necessary to extend it by 0·5 mm?

(g) What is the wavelength in air at 0 °C of a sound wave of frequency 256 Hz?

(h) What is the average (rms) velocity of a hydrogen molecule at 27 °C?

(i) Estimate a value of G from g.

(j) Parallel monochromatic red light is incident normally on a diffraction grating. (i) If the first order spectrum is at an angle of 30° to the normal, estimate the number of lines per mm of the grating. What roughly would be the angle for (ii) green light and (iii) violet light?

(k) How much energy is radiated per second by a black sphere of radius 20 mm at a temperature of 500 K?

(l) What is (i) the velocity and (ii) the wavelength of an electron whose energy is 1 keV?

(m) What magnetic field is required to bend a stream of protons of energy 1 MeV into a circle of radius 500 mm?

(n) What is the wavelength of light that will just ionize a hydrogen atom in the ground state?

(o) What is the energy equivalent in MeV of a mass of unit atomic weight?

12.2 Calculate the following in your head:

(a) $1{\cdot}00025 \times 1{\cdot}00041 \times 0{\cdot}99987$,

(b) $912{\cdot}64 \times \left(\dfrac{7200{\cdot}0}{7200{\cdot}9}\right)$,

(c) $(9{\cdot}100)^{\frac{1}{2}}$.

13

Writing a paper

13.1 Introduction

The communication of ideas, theories, and experimental results is an important part of scientific work. Vast quantities of scientific literature are pouring out into the world, and if you take up a scientific career of any kind you are almost certain to add to the flood. If you can achieve a good standard of writing, two benefits will accrue - one to yourself when people take note of what you have to say, and the other to the rest of the world who - strange to say - prefer their reading matter to be clear and interesting rather than obscure and dull.

We are going to consider some elementary features of good scientific writing in the present chapter. To make the discussion specific we shall confine it to a paper on some experimental work in physics, but much of what we have to say applies to scientific writing in general.

13.2 Title

The title serves to identify the paper. It should be brief - not more than about 10 words. You should bear in mind that the title will ultimately appear in a subject index. The compiler of an index relies heavily on the words in the title in deciding where it should appear. So if there are one or two key words which help to classify the work, try to put them in the title.

13.3 Abstract

Every paper should have an abstract of about 100 words or so, giving positive information about its contents.

The abstract serves two classes of reader. It enables those who work in the subject to decide whether they want to read the paper; and it provides a summary for those who have only a general interest in the subject - they can obtain the essential results without having to read the whole paper. The abstract should therefore not only indicate the general scope of the paper but should contain the final numerical results and the main conclusions.

13.4 Plan of paper

Most papers - unless they are very short - are divided into sections. The following is a fairly common division:

Introduction
Experimental method
Results
Discussion

Some papers describing experimental work also contain theoretical material which might well constitute an additional section, coming after either Introduction or Results.

Although the actual plan of a paper depends to some extent on its contents, you can see that the one above is logical and you should try to follow it, at least in a general way. We consider each section in turn.

13.5 Sections of paper

(*a*) *Introduction*. The Introduction is an important part of the paper. Most experiments are part of a general investigation of a physical problem. The Introduction should make clear

(i) the physical interest of the problem,
(ii) the part played by the experiment in the general investigation,
(iii) the relation of the experiment to any previous work.

These points add up to your answering the question 'Why did you do the experiment or what was its *object*?'

You may assume that the reader of the main body of the paper has a certain background knowledge of the subject, but it may be that someone starting the paper does not have this knowledge. The Introduction should serve as a possible starting point for him. You may not wish to go back to the beginning of the subject, in which case you should give references - not too many - to published work which does provide the necessary background. The Introduction plus this work should bring the reader to the point where he is ready to hear about your experiment.

We give an example of a splendid opening. It is from J. J. Thomson's paper on Cathode Rays (Thomson 1897), effectively announcing the discovery of the electron.

CATHODE RAYS

The experiments discussed in this paper were undertaken in the hope of gaining some information as to the nature of the Cathode Rays. The most diverse opinions are held as to these rays; according to the almost unanimous opinion of German physicists they are due to some process in the aether to which - inasmuch as in a uniform magnetic field their course is circular and not rectilinear - no phenomenon hitherto observed is analogous: another view of these rays is that, so far from being wholly aetherial, they are in fact wholly material, and that they mark the paths of particles of matter charged with negative electricity. It would seem at first sight that it ought not to be difficult to discriminate between views so different, yet experience shows that this is not the case, as amongst the physicists who have most deeply studied the subject can be found supporters of either theory.

The electrified-particle theory has for purposes of research a great advantage over the aetherial theory, since it is definite and its consequences can be predicted; with the aetherial theory it is impossible to predict what will happen under any given circumstances, as on this theory we are dealing with hitherto unobserved phenomena in the aether, of whose laws we are ignorant.

The following experiments were made to test some of the consequences of the electrified-particle theory.

We shall have more to say about this passage later. But in the meantime, notice how clearly and directly Thomson has given the kind of information that should appear in the Introduction - the opening sentence is a model of its kind.

(*b*) *Experimental method.* In this section comes the description of the apparatus. The amount of detail here varies considerably, and you must use your own judgement, but a few general principles may serve as a guide.

If the apparatus you used is of a standard kind, it is probably sufficient to say what it was and give a reference so that anyone interested can find a full description. On the other hand, if it contains some novel features, they should be described in some detail. If the paper is intended for a journal devoted to descriptions of instruments and apparatus, such as the *Journal of Physics E*, still more detail would be appropriate. But we shall suppose that this is not the case and that the main interest is in the results and their interpretation, rather than on the apparatus used.

Although you may assume that the reader of this section has a certain familiarity with the background to the work, you should not go farther than this. You should certainly not aim the paper directly at other experimenters using the same or similar apparatus. So you should not use esoteric phrases understood only by such workers. Nor should you include minute experimental details of interest only to them.

(*c*) *Results*. In general it is neither possible nor desirable to give all the measurements. They would only confuse and distract the reader. He would have to spend time assessing their relative importance and extracting the essential results. But that is your job before writing the paper. You should, therefore, give only

(i) a representative sample of some of the basic measurements,
(ii) the important results.

Note the word *representative* in (i). The sample that you present in the paper should give a faithful picture of the quality, precision, and reproducibility of the measurements. So if you have fifty sets of them, you do not reproduce the second best with the caption 'Typical set'.

(*d*) *Discussion*. The heading speaks for itself. Like the Introduction, this section is an important part of the paper. It should include

(i) comparison with other similar measurements, if there are any,
(ii) comparison with relevant theories,
(iii) discussion of the state of the problem under investigation in the light of your results. This is the *conclusion*, the counterpart to the *object* of the experiment, given in the Introduction.

13.6 Diagrams, graphs, and tables

Almost everything said in chapters 10 and 11 about diagrams, graphs, and tables applies to their use in papers.

A diagram can be a great help in understanding the text. Unless the apparatus is completely standard, a diagram of it should nearly always be included. Graphs are a very convenient and common way of displaying the results. They should be kept simple; the same applies to diagrams. In the final printed version, graphs and diagrams are usually reduced in scale by a factor of two or three, so unless they are bold and clear in the original, they will look fussy and complicated in the final version.

Tables are a very good way of presenting results. They have the big advantage that they stand out, and the reader can find the results easily.

13.7 Instructions to authors

Most scientific journals produce a pamphlet with instructions to authors, so that their papers may conform to the general style of the journal. You are advised to read this *before* the paper is typed in its final form; otherwise someone - probably you - will have to spend a lot of time bringing it into line later on.

The pamphlet gives instructions on the form of sectional headings, abbreviations, references, footnotes, tables, diagrams and graphs. It also gives instructions on the arrangement of mathematical material. This is important, because the way you write a mathematical expression in longhand may be inconvenient or expensive to set up in print. For example, most journals prefer

$$(a^2+b^2)^{1/2} \quad \text{to} \quad \sqrt{a^2+b^2}.$$

If the journal does not produce a pamphlet, you should examine a recent issue.

13.8 Clarity

Clarity is an essential quality in scientific writing. We may distinguish two kinds.

(a) Structural clarity. Writing may be said to have structural clarity when the reader can readily follow the outline of the argument - or see the wood despite the trees. Similar topics are grouped together, and the groups arranged in a logical order.

You are strongly advised to construct a framework before writing the paper. This is a skeleton outline in which all the ideas, arguments, experimental details and so on are represented by a word or phrase. When the items are in this form, the arrangement is seen much more clearly and, moreover, is easily changed if not satisfactory. The main sections of the scheme should correpond to the plan given in section 13.4.

(b) Expositional clarity. The other type of clarity, which may be called expositional clarity, is making the reader understand exactly what you are trying to tell him at each stage in the discussion.

Look at the extract from the paper on Cathode Rays again. It is crystal clear. We are led firmly from one point to the next. Notice the phrase 'so far from being wholly aetherial'. It could have been omitted, and we would still have followed the discussion; but the explicit contrast is helpful. Making it easy for the reader is a worthwhile object in any writing, but particularly in scientific writing.

Perhaps you feel that the example we have given is not a very severe test for the writer, because he is explaining something simple. That is true, but the reason it is simple is that Thomson has made it so. He has *selected* the important features of the theories of the nature of the rays. He is able to do this because he *understands the physics.* And this a fundamental point. Clear writing depends on clear thinking. You will not be able to produce a clear and logical paper unless you do understand the physics.

13.9 Good English

We come to the final link in the chain between you and the reader - the words themselves. Good English in scientific writing is not just a matter of correct grammar - though that is not unimportant - it is choosing words and composing sentences to say exactly what you mean as concisely and pleasantly as possible. Some instructive books on the subject are given on p. 205. We make a few miscellaneous points here.

(a) Students are often discouraged from using 'I' in their accounts. There seems no sensible reason for this. When you are describing an experiment you actually did, the 'I' style is a natural one and enables you to use the active voice, which is simpler and more direct than the passive. However, it must be admitted that nowadays few papers, even those describing experiments, are written in the first person. So if you want to be conventional, avoid it. But if you do use it, you will be in the company of Newton, Faraday, Maxwell, and Thomson, which, as Damon Runyon might have said, is by no means bad company at that.

(b) On the whole, short sentences make for clarity, but variety is necessary to avoid monotony. You can be clear in long sentences - Thomson's are hardly short - but it takes more skill.

(c) Paragraphs can help the reader to follow the argument. Start a new paragraph when you are starting a fresh point, or when you start discussing a point from a different angle.

(d) Avoid verbiage, roundabout ways of saying things, and redundant adverbs. Thus the second version should be preferred to the first in the following examples:

(i) Calculations were carried out on the basis of a comparatively rough approximation.

(ii) Approximate calculations were made.

(i) Similar considerations may be applied in the case of copper with a view to testing to what extent the theory is capable of providing a correct estimate of the elastic properties of this metal.

(ii) The elastic properties of copper may be calculated in the same way as a further test of the theory.

(e) Avoid qualifying a noun with a long string of adjectives - most of which are not adjectives anyhow. For example

The time-of-flight inelastic thermal neutron scattering apparatus ... should be replaced by

The time-of-flight apparatus for the inelastic scattering of thermal neutrons ....

(f) The unattached participle is a common fault in scientific writing. Sentences like

Inserting equation (3), the expression becomes ...

or

Using a multimeter, the voltage was found to be ...

occur frequently, so much so that you may not even realize what is wrong with them. Perhaps the following example taken from Fowler will show. A firm wrote to a customer 'Dear Sir, Being desirous of clearing our books, will you please send a cheque to settle your account. Yours etc.' They received the following reply: 'Sirs, you have been misinformed. I have no desire to clear your books.'

13.10 Conclusion

Not everyone can write great literature, but anyone can write good, clear English - if he is prepared to take the trouble. Be critical of what you write. Ask yourself continually whether what you have written is logical, clear, concise. If not, try again - and again. Hard writing makes easy reading. Give your work to others to read for critical comment; read and criticize theirs.

You should not regard good experimenting and good writing as separate things. There is beauty in both, and it is no accident that the greatest scientists like Galileo and Newton have produced some of the finest scientific writing. But we shall allow a non-scientist - Cervantes - the final word.

> 'Study to explain your Thoughts and set them in the truest Light, labouring as much as possible not to leave them dark nor intricate, but clear and intelligible.'

APPENDIX A

Values of the Gaussian function and the Gaussian integral function

$$f(z)=\frac{1}{\sqrt{(2\pi)}}\exp(-z^2/2) \qquad \phi(z)=\sqrt{\left(\frac{2}{\pi}\right)}\int_0^z \exp(-t^2/2)\,\mathrm{d}t$$

z	$f(z)$	$\phi(z)$	z	$f(z)$	$\phi(z)$
0·0	0·399	0·000	1·5	0·130	0·866
0·1	0·397	0·080	1·6	0·111	0·890
0·2	0·391	0·159	1·7	0·094	0·911
0·3	0·381	0·236	1·8	0·079	0·928
0·4	0·368	0·311	1·9	0·066	0·943
0·5	0·352	0·383	2·0	0·054	0·954
0·6	0·333	0·451	2·2	0·035	0·972
0·7	0·312	0·516	2·4	0·022	0·984
0·8	0·290	0·576	2·6	0·014	0·991
0·9	0·266	0·632	2·8	0·008	0·995
1·0	0·242	0·683	3·0	0·0044	0·9973
1·1	0·218	0·729	3·5	0·0009	0·9995
1·2	0·194	0·770	4·0	0·0001	0·99994
1·3	0·171	0·806			
1·4	0·150	0·838			

APPENDIX B

Evaluation of some integrals connected with the Gaussian function

B.1 $\int_{-\infty}^{\infty} \exp(-x^2)\, dx$

First consider the integral with finite limits. We put

$$U = \int_{-a}^{a} \exp(-x^2)\, dx = \int_{-a}^{a} \exp(-y^2)\, dy, \tag{B.1}$$

since the variable is a dummy. Therefore

$$U^2 = \int_{-a}^{a} \exp(-x^2)\, dx \int_{-a}^{a} \exp(-y^2)\, dy. \tag{B.2}$$

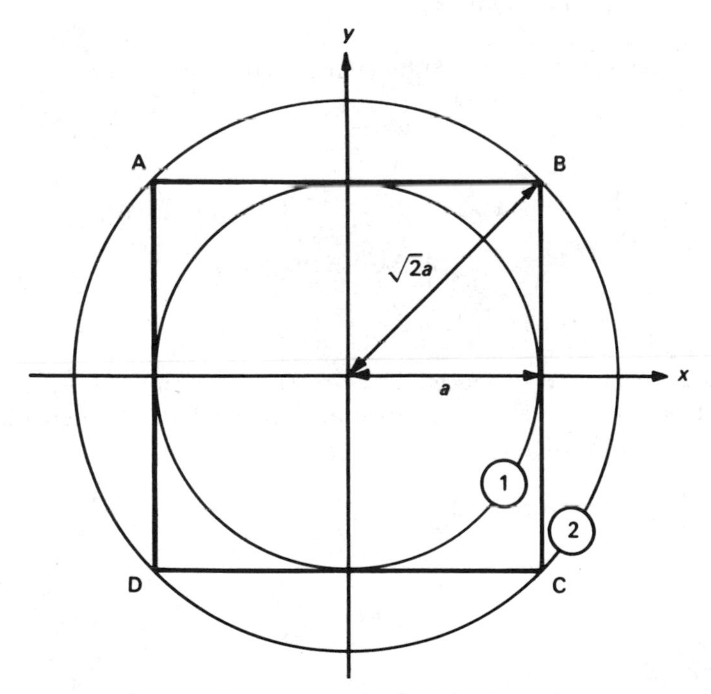

Fig. B.1. Evaluation of the Gaussian integral.

Since one integrand is a function only of x and the other only of y, the right-hand side is equal to $\exp[-(x^2+y^2)]$ integrated over the square ABCD in Fig. B.1, i.e.

$$U^2 = \int_{-a}^{a}\int_{-a}^{a} \exp[-(x^2+y^2)]\,dx\,dy. \qquad (B.3)$$

We now change to polar coordinates r, θ. The area element $dx\,dy$ is replaced by $r\,dr\,d\theta$, and $\exp[-(x^2+y^2)]$ becomes $\exp(-r^2)$.

Consider the integral of $\exp(-r^2)$ over a circle of radius b

$$W(b) = \int_0^b\int_0^{2\pi} \exp(-r^2)\,r\,dr\,d\theta = \pi[1-\exp(-b^2)]. \qquad (B.4)$$

Clearly U^2 is greater than $W(b)$ for $b=a$ (circle ① in Fig. B.1) but less than $W(b)$ for $b=\sqrt{2}a$ (circle ②); i.e. U^2 lies between

$$\pi[1-\exp(-a^2)] \quad \text{and} \quad \pi[1-\exp(-2a^2)].$$

But as a tends to infinity, both these limits tend to π, and therefore U^2 also tends to π. So

$$\int_{-\infty}^{\infty} \exp(-x^2)\,dx = \sqrt{\pi}. \qquad (B.5)$$

B.2 $I_0 = \int_{-\infty}^{\infty} \exp(-x^2/2\sigma^2)\,dx$

This is evaluated from (B.5) by changing the variable to

$$y = \frac{x}{\sqrt{2}\sigma}. \qquad (B.6)$$

$$\int_{-\infty}^{\infty} \exp(-x^2/2\sigma^2)\,dx = \sqrt{2}\sigma \int_{-\infty}^{\infty} \exp(-y^2)\,dy. \qquad (B.7)$$

$$I_0 = \sqrt{(2\pi)}\sigma. \qquad (B.8)$$

B.3 $I_n = \int_{-\infty}^{\infty} x^n \exp(-x^2/2\sigma^2)\,dx$

n is a positive, even integer or zero. Having evaluated I_0, we may evaluate any I_n by treating σ as a variable parameter and differentiating under the integral sign.

$$\frac{d}{d\sigma}\int_{-\infty}^{\infty} x^n \exp(-x^2/2\sigma^2)\,dx = \frac{1}{\sigma^3}\int_{-\infty}^{\infty} x^{n+2} \exp(-x^2/2\sigma^2)\,dx, \qquad (B.9)$$

i.e.

$$I_{n+2} = \sigma^3 \frac{dI_n}{d\sigma}. \qquad (B.10)$$

So for $n \geqslant 2$,

$$I_n = 1\,.\,3\,.\,5\ldots(n-1)\sqrt{(2\pi)}\sigma^{n+1}. \qquad (B.11)$$

APPENDIX C

The variance of s^2 for a Gaussian distribution

We prove the result, stated in section 3.7, that for a Gaussian distribution the variance of s^2 is $2\langle s^2\rangle^2/(n-1)$.

For one set of n readings

$$s^2 = \frac{1}{n}\sum d_i^2. \tag{C.1}$$

The error in s^2 is

$$u = s^2 - \langle s^2\rangle. \tag{C.2}$$

The quantity we require is

$$\begin{aligned}\langle u^2\rangle &= \langle s^4 - 2s^2\langle s^2\rangle + \langle s^2\rangle^2\rangle \\ &= \langle s^4\rangle - \langle s^2\rangle^2.\end{aligned} \tag{C.3}$$

From (3.8) and (3.16)

$$\begin{aligned}s^2 &= \frac{1}{n}\sum e_i^2 - \frac{1}{n^2}\left(\sum e_i\right)^2 \\ &= \left(\frac{1}{n} - \frac{1}{n^2}\right)\sum e_i^2 - \frac{1}{n^2}\sum_{\substack{i\ j \\ i\neq j}}\sum e_i e_j.\end{aligned} \tag{C.4}$$

Square both sides and average over the distribution. The average of any term containing an odd power of e is zero. Therefore

$$\langle s^2\rangle = \left(1 - \frac{1}{n}\right)\langle e^2\rangle, \tag{C.5}$$

and

$$\langle s^4\rangle = \left(\frac{1}{n} - \frac{1}{n^2}\right)^2 n\langle e^4\rangle + \left(\frac{1}{n} - \frac{1}{n^2}\right)^2 n(n-1)\langle e^2\rangle^2 + \frac{1}{n^4}2n(n-1)\langle e^2\rangle^2. \tag{C.6}$$

From (C.3), (C.5), and (C.6)

$$\frac{\langle u^2\rangle}{\langle s^2\rangle^2} = \frac{1}{n}\left[\frac{\langle e^4\rangle}{\langle e^2\rangle^2} - \frac{n-3}{n-1}\right]. \tag{C.7}$$

For a Gaussian distribution we have from Table 3.3

$$\langle e^4 \rangle = 3\sigma^4 \quad \text{and} \quad \langle e^2 \rangle = \sigma^2, \tag{C.8}$$

whence

$$\langle u^2 \rangle = \frac{2}{n-1} \langle s^2 \rangle^2. \tag{C.9}$$

APPENDIX D

The binomial and Poisson distributions

The Poisson distribution may be regarded as a limiting case of the binomial distribution, so we consider the latter first.

D.1 Binomial distribution

(a) Derivation. Suppose an event can have only two outcomes which we denote by A and B. Let the probability of outcome A be p. Then the probability of outcome B is $q = 1 - p$. If the event occurs N times, the probability $w_N(n)$ that A comes up n times and B $N - n$ times is equal to the number of ways of selecting n objects from N, i.e. ${}_NC_n$, times the probability that the first n events give A and the remaining $N - n$ events give B. Therefore

$$w_N(n) = \frac{N!}{n!(N-n)!} p^n q^{N-n}. \tag{D.1}$$

This probability function is known as the binomial distribution. It is specified by giving the values of the two parameters N and p. Figure D.1 shows the distribution for $N = 10$, $p = \frac{1}{3}$.

If we sum $w_N(n)$ from $n = 0$ to $n = N$, the result must be unity. To check that this is so we use a mathematical trick, which will also be useful later. We define a function $g(z)$ by

$$g(z) = (q + zp)^N. \tag{D.2}$$

Then the coefficient of z^n is just $w_N(n)$, thus

$$\begin{aligned} g(z) - (q + zp)^N &= q^N + zNpq^{N-1} + z^2 \frac{N(N-1)}{2!} p^2 q^{N-2} + \ldots + z^N p^N \\ &= w_N(0) + z w_N(1) + z^2 w_N(2) + \ldots + z^N w_N(N). \end{aligned} \tag{D.3}$$

$\sum_{n=0}^{N} w_N(n)$ is obtained from the last line by putting $z = 1$. Therefore

$$\sum_{n=0}^{N} w_N(n) = g(1) = (q + p)^N = 1, \quad \text{since} \quad q + p = 1. \tag{D.4}$$

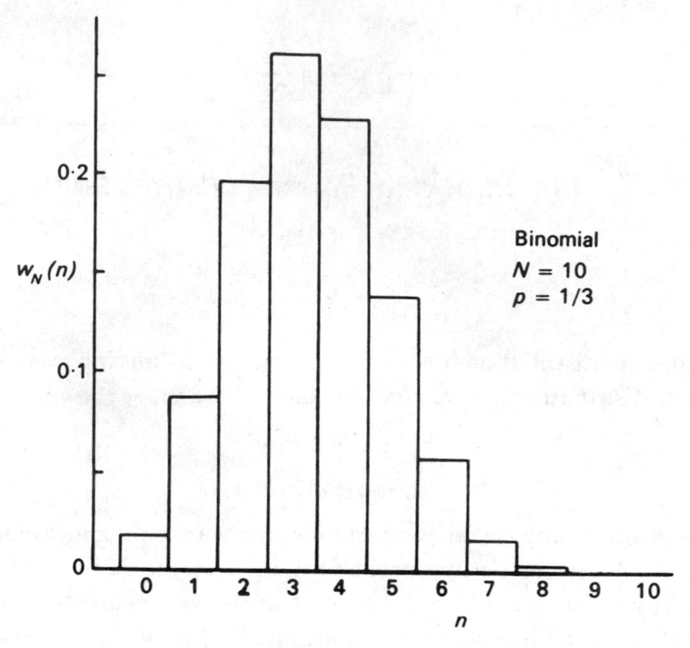

Fig. D.1. The binomial distribution for $N=10$, $p=\frac{1}{3}$.

(*b*) *Mean value of n*. The mean value of n for the distribution is

$$\langle n\rangle = \sum_{n=0}^{N} n w_N(n) = \left(\frac{\mathrm{d}g}{\mathrm{d}z}\right)_{z=1} \qquad \text{(from D.3)}$$

$$= Np(q+p)^{N-1}$$

$$= Np. \tag{D.5}$$

We might have guessed this result. If the probability of A coming up in a single event is p, we would expect the average number of As in N events to be Np.

(*c*) *Standard deviation*. The standard deviation σ is given by

$$\sigma^2 = \langle (n-\langle n\rangle)^2\rangle = \langle n^2 - 2n\langle n\rangle + \langle n\rangle^2\rangle$$

$$= \langle n(n-1)\rangle + \langle n\rangle - \langle n\rangle^2. \tag{D.6}$$

Now

$$\langle n(n-1)\rangle = \sum_{n=0}^{N} n(n-1) w_N(n) = \left(\frac{\mathrm{d}^2 g}{\mathrm{d}z^2}\right)_{z=1}$$

$$= N(N-1)p^2(q+p)^{N-2}$$

$$= N(N-1)p^2. \tag{D.7}$$

From (D.5), (D.6), and (D.7)

$$\begin{aligned}\sigma^2 &= N(N-1)p^2 + Np - N^2p^2 \\ &= Np(1-p), \qquad\qquad (D.8)\end{aligned}$$

or

$$\sigma = \surd(Npq). \qquad\qquad (D.9)$$

D.2 Poisson distribution

(a) Derivation. The Poisson distribution is the limiting case of the binomial distribution when N tends to infinity and p tends to zero, in such a way that the product Np is equal to a finite constant which we denote by a. We require an expression for the probability $w_a(n)$ that event A occurs n times.

$$\begin{aligned}w_a(n) &= \operatorname*{limit}_{\substack{N\to\infty \\ p\to 0 \\ Np=a}} \frac{N!}{n!(N-n)!} p^n q^{N-n} \\ &= \frac{a^n}{n!} \operatorname{limit} \frac{N(N-1)\ldots(N-n+1)}{N^n} q^{N-n}. \qquad\qquad (D.10)\end{aligned}$$

As N tends to infinity, the quantity

$$\frac{N(N-1)\ldots(N-n+1)}{N^n}$$

tends to unity. (Remember that n is finite, and therefore small compared with N.)

Also

$$\begin{aligned}q^{N-n} &= (1-p)^{N-n} = \left(1-\frac{a}{N}\right)^{N-n} \\ &= \frac{\left(1-\frac{a}{N}\right)^N}{\left(1-\frac{a}{N}\right)^n}. \qquad\qquad (D.11)\end{aligned}$$

As N tends to infinity the numerator tends to $\exp(-a)$ and the denominator to unity.

Collecting these results we have

$$w_a(n) = \exp(-a)\frac{a^n}{n!}, \qquad\qquad (D.12)$$

which is the Poisson distribution. Notice that, whereas the binomial distribution is specified by two parameters, the Poisson distribution is specified by the single parameter a.

We can check that summing $w_a(n)$ over n gives unity.

$$\sum_{n=0}^{\infty} w_a(n) = \exp(-a) \sum_{n=0}^{\infty} \frac{a^n}{n!}$$

$$= \exp(-a) \exp a = 1. \tag{D.13}$$

(b) Mean value of n and standard deviation. We may use the results of the binomial distribution to obtain the mean value of n and the standard deviation.

$$\langle n \rangle = Np = a. \tag{D.14}$$

$$\sigma = \sqrt{(Npq)} = \sqrt{a}, \tag{D.15}$$

since $q = 1$ in the limit. As a becomes large, the Poisson distribution becomes more and more symmetric, and the points tend to lie on a Gaussian with mean a and standard deviation $\sqrt{a}$.

(c) Application. The Poisson distribution applies to the counting of particles when the average rate of arrival is constant - a common situation in atomic and nuclear physics. Suppose we are counting electrons with a scintillation counter, and we record the numbers that arrive in successive periods of, say, 10 seconds. The numbers form a Poisson distribution.

To see how the situation relates to our definition of the Poisson distribution, imagine that the 10-second interval is divided into N sub-intervals, where N is very large - say 10^8. Suppose that a, the average

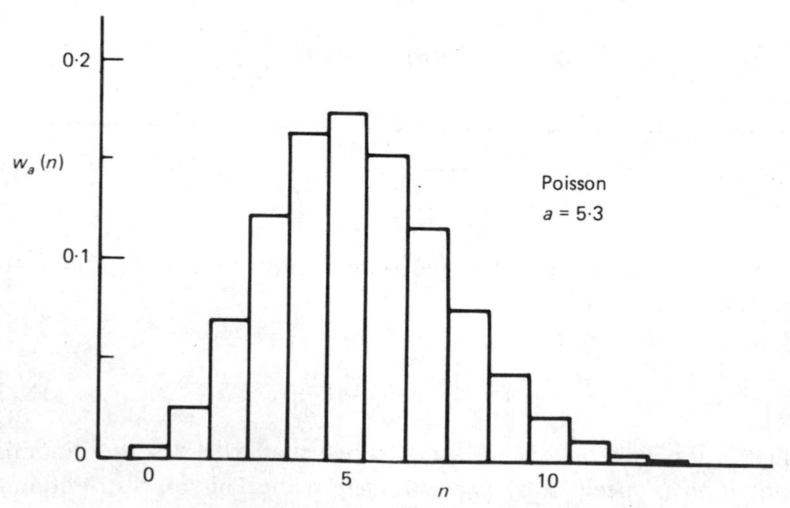

Fig. D.2. The Poisson distribution for $a = 5{\cdot}3$.

number of electrons recorded in the 10-second intervals, is 5·3. Then the probability of an electron arriving in any particular sub-interval is

$$p = \frac{a}{N} = 5{\cdot}3 \times 10^{-8}. \qquad \text{(D.16)}$$

Outcome A in this situation is the arrival of an electron in the sub-interval; outcome B is the non-arrival of an electron. (p is so small that we ignore the possibility of two electrons arriving in the same sub-interval.)

The probability of n electrons arriving in a 10-second interval is

$$w_{5\cdot3}(n) = \exp(-5{\cdot}3)\frac{(5{\cdot}3)^n}{n!}. \qquad \text{(D.17)}$$

The standard deviation is $\sqrt{5{\cdot}3}$. The Poisson distribution for $a = 5{\cdot}3$ is shown in Fig. D.2.

APPENDIX E

The straight line – the standard error in the slope and intercept

In section 4.2 we used the method of least squares to calculate the best values of m and c in the equation $y = mx + c$, for a set of measurements consisting of n pairs $(x_1, y_1), (x_2, y_2), \ldots, (x_n, y_n)$ with equal weights. We here derive the expressions for the standard errors Δm and Δc given in formulae (4.31) and (4.32).

Imagine that we keep repeating the measurements so that we have many sets each of n pairs, the measurements being made in such a way that the values $x_1, x_2, \ldots, x_n$ are the same for all the sets. In other words, if we look at the first pair x_1, y_1 in all the sets, x_1 is the same throughout, but the value of y_1 varies. We shall have a distribution of y_1s centred about Y_1, the true value of y for x_1. Similarly for all the second pairs; x_2 is the same in each case, but the y_2s vary and form a distribution centred on Y_2, the true value of y for x_2. And so on for all the pairs. Since the n pairs have equal weights, the standard errors of the n distributions are equal and we denote them by σ. The situation is shown schematically in Fig. E.1. We assume that for each set there is no correlation between the errors in two different ys.

Now the x_i, Y_i values are related by

$$Y_i = Mx_i + C. \tag{E.1}$$

This is the true line, and M and C are the true values of the slope and intercept.

For each set of $n(x_i, y_i)$ pairs, we can calculate the values of m and c from the expressions given in (4.25) and (4.26). The value of m averaged over all the sets is M, and the standard error in a single value is Δm, where

$$(\Delta m)^2 = \langle (m - M)^2 \rangle, \tag{E.2}$$

the average again being taken over all the sets. Similarly the average value of c is C, and the standard error in c is Δc, where

$$(\Delta c)^2 = \langle (c - C)^2 \rangle. \tag{E.3}$$

In an actual experiment we only have one set of $n(x_i, y_i)$ values. The

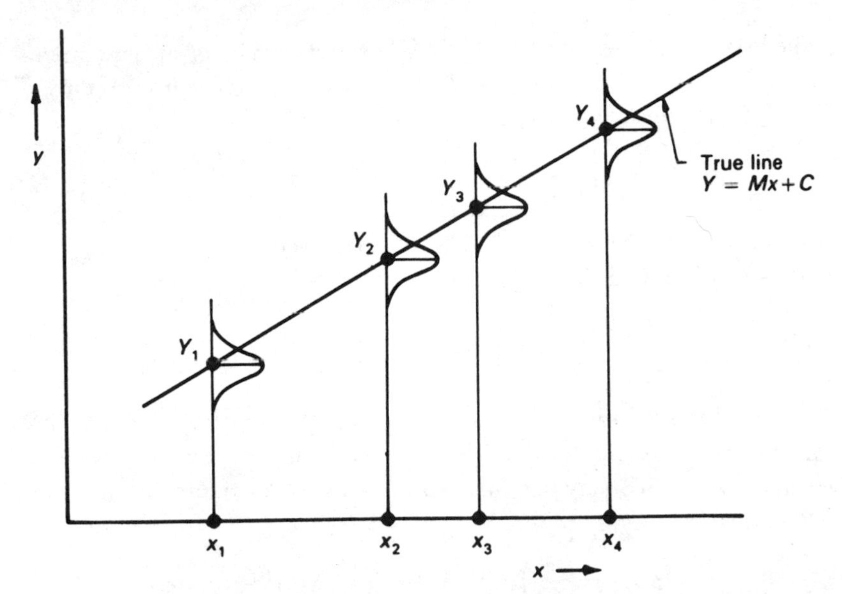

Fig. E.1. Repetition of measurements with a fixed set of x values. For each x the values of y form a distribution centred about the true value Y.

values for m and c for this particular set are our best estimates of M and C. The problem is to obtain estimates for Δm and Δc.

The algebra is considerably simplified if we change the independent variable from x to ξ, given by

$$\xi = x - \bar{x}, \tag{E.4}$$

where

$$\bar{x} = \frac{1}{n}\sum x_i. \tag{E.5}$$

Clearly

$$\sum \xi_i = \sum (x_i - \bar{x}) = 0. \tag{E.6}$$

The quantity D is defined by

$$D = \sum \xi_i^2 = \sum (x_i - \bar{x})^2 = \sum x_i^2 - n\bar{x}^2. \tag{E.7}$$

The line

$$y = mx + c \tag{E.8}$$

becomes

$$y = m(\xi + \bar{x}) + c \tag{E.9}$$

$$= m\xi + b. \tag{E.10}$$

where

$$b = m\bar{x} + c. \tag{E.11}$$

The best values of m and b for a given set of n pairs of measurements are obtained from (4.25) and (4.26) with c replaced by b and x by ξ. Since $\sum \xi_i = 0$, these equations become

$$m = \frac{1}{D} \sum \xi_i y_i, \qquad b = \bar{y} = \frac{1}{n} \sum y_i. \tag{E.12}$$

From now on the symbols m and b refer to these values.

We see that m is a linear function of the y_i. For a given set

$$m = \frac{\xi_1}{D} y_1 + \frac{\xi_2}{D} y_2 + \ldots \tag{E.13}$$

The coefficients of the ys are the same for all the sets. Since we are assuming that in each set there is no correlation between the errors in two different ys, we may use (4.17) and (4.18) to calculate Δm in terms of the errors in the ys.

$$(\Delta m)^2 = \left(\frac{\xi_1}{D}\right)^2 (\Delta y_1)^2 + \left(\frac{\xi_2}{D}\right)^2 (\Delta y_2)^2 + \ldots \tag{E.14}$$

But

$$(\Delta y_1)^2 = (\Delta y_2)^2 = \ldots = \sigma^2. \tag{E.15}$$

So we have

$$(\Delta m)^2 = \frac{\sum \xi_i^2}{D^2} \sigma^2 = \frac{\sigma^2}{D}. \tag{E.16}$$

Similarly,

$$(\Delta b)^2 = \frac{1}{n} \sigma^2. \tag{E.17}$$

We actually require $(\Delta c)^2$, which, from (E.11), is given by

$$(\Delta c)^2 = (\Delta b)^2 + \bar{x}^2 (\Delta m)^2 \tag{E.18}$$

$$= \left(\frac{1}{n} + \frac{\bar{x}^2}{D}\right) \sigma^2. \tag{E.19}$$

(See comment at the end of the Appendix.)

The estimate of σ is obtained as follows. If B is the true value of b, then

$$Y_i = M\xi_i + B. \tag{E.20}$$

Adding these equations for each i gives an expression for B (since $\sum \xi_i = 0$). Similarly multiplying each one by ξ_i and then adding gives an expression for M. Thus

$$M = \frac{1}{D} \sum \xi_i Y_i, \qquad B = \frac{1}{n} \sum Y_i. \tag{E.21}$$

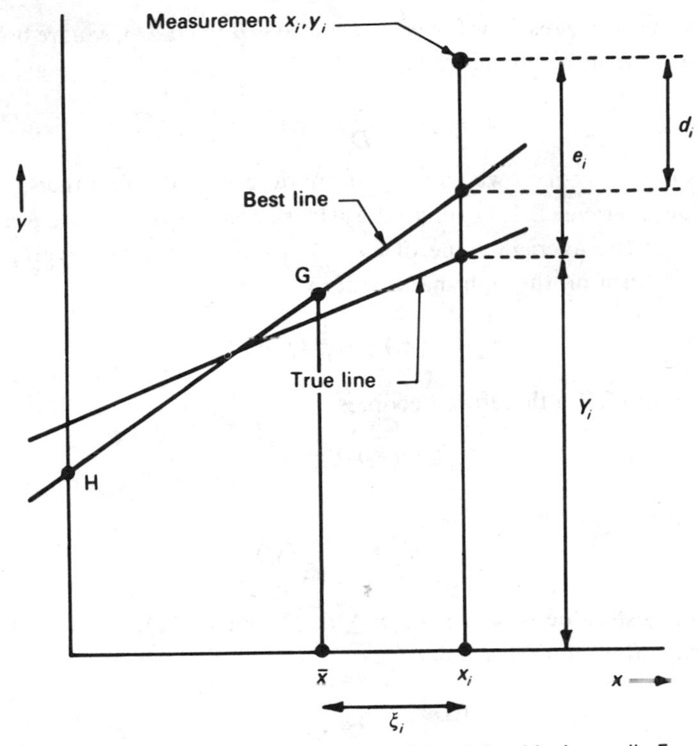

Fig. E.2. Diagram showing various quantities defined in Appendix E.

The error in the ith y reading is

$$e_i = y_i - Y_i = y_i - (M\xi_i + B). \tag{E.22}$$

At the point ξ_i, the best line gives $m\xi_i + b$ for the value of y. The residual d_i is therefore

$$d_i = y_i - (m\xi_i + b) \tag{E.23}$$

- Fig. E.2. As in the case of a single variable, the errors e_i are not known, but the residuals d_i are known. The root-mean-square value of d_i for the n points is denoted by s as before.

From (E.22) and (E.23)

$$d_i = e_i - [(m - M)\xi_i + (b - B)]. \tag{E.24}$$

From (E.12) and (E.21)

$$m - M = \frac{1}{D}\sum \xi_i(y_i - Y_i) = \frac{1}{D}\sum \xi_i e_i, \tag{E.25}$$

$$b - B = \frac{1}{n}\sum e_i. \tag{E.26}$$

Insert these expressions for $m-M$ and $b-B$ in (E.24), square both sides, and sum over i. This gives

$$\sum d_i^2 = \sum e_i^2 - \frac{1}{D}(\sum \xi_i e_i)^2 - \frac{1}{n}(\sum e_i)^2. \tag{E.27}$$

(In summing over i we have again made use of the fact that $\sum \xi_i = 0$.)

Now average (E.27) over all the sets, remembering that the ξ_i are fixed, and that the average value of $e_i e_j$ for $i \neq j$ is zero. The average of the middle term on the right-hand side is

$$\frac{1}{D}\langle(\sum \xi_i e_i)^2\rangle = \frac{1}{D}\langle \sum \xi_i^2 e_i^2 \rangle = \sigma^2. \tag{E.28}$$

Equation (E.27) therefore becomes

$$n\langle s^2\rangle = n\sigma^2 - \sigma^2 - \sigma^2, \tag{E.29}$$

or

$$\sigma^2 = \frac{n}{n-2}\langle s^2 \rangle. \tag{E.30}$$

Our best value of $\langle s^2\rangle$ is $(1/n)\sum d_i^2$. From (E.16), (E.19), and (E.30) we have the required results

$$(\Delta m)^2 \approx \frac{1}{D}\frac{\sum d_i^2}{n-2}, \tag{E.31}$$

$$(\Delta c)^2 \approx \left(\frac{1}{n} + \frac{\bar{x}^2}{D}\right)\frac{\sum d_i^2}{n-2}. \tag{E.32}$$

The generalization to the case of unequal weights is readily made. If the ith point has weight w_i, the variance of the ith distribution is put equal to $\sigma^2(\sum w_i)/nw_i$, where σ is a constant. The results given on p. 44 then follow by reasoning closely similar to that above.

Comment on the dependence of m, c, and b

Equation (E.18) assumes that the values of m and b are independent, which may readily be proved by calculating $(m-M)(b-B)$ from (E.12) and (E.21). The average value of this quantity is seen to be zero. However, m and c are *not* independent, for

$$\langle (m-M)(c-C)\rangle = -\bar{x}(\Delta m)^2. \tag{E.33}$$

Since m and $b(=\bar{y})$ are independent, while m and c are not, the best line should be written as

$$y = (m \pm \Delta m)(x - \bar{x}) + b \pm \Delta b, \tag{E.34}$$

and not as

$$y = (m \pm \Delta m)x + c \pm \Delta c. \tag{E.35}$$

Equation (E.34) implies correctly that the error Δy in the best value of y at any value of x is given by

$$(\Delta y)^2 = (\Delta b)^2 + (x - \bar{x})^2(\Delta m)^2. \tag{E.36}$$

The best line may therefore be regarded as pivoting about the centre of gravity of the points - G in Fig. E.2. The errors in the y value of the pivot and in the slope of the line contribute independently to Δy. Equation (E.35) implies incorrectly that the pivot point is H.

APPENDIX F

SI units

The system of units used throughout this book is known as SI, an abbreviation for Système International d'Unités. It is a comprehensive, logical system, designed for use in all branches of science and technology. It was formally approved in 1960 by the General Conference of Weights and Measures, the international organization responsible for maintaining standards of measurement. Apart from its intrinsic merits, it has the great advantage that *one* system covers all situations - theoretical and practical.

The following are the essential features of the system:

(1) SI is a metric system. There are seven base units (see next section), the metre and kilogram replacing the centimetre and gram of the old c.g.s. system.

(2) The derived units are directly related to the base units. For example, the unit of acceleration is $1\ \mathrm{m\ s^{-2}}$. The unit of force is the newton, which is the force required to give a body of mass 1 kg an acceleration of $1\ \mathrm{m\ s^{-2}}$. The unit of energy is the joule, which is the work done when a force of 1 N moves a body a distance of 1 m.

The use of auxiliary units is discouraged in SI. Thus the unit of pressure, the pascal, is $1\ \mathrm{N\ m^{-2}}$; the atmosphere and the torr are not used. Similarly the calorie is not used; all forms of energy are measured in joules. (However, the electron volt remains - at least for the time being.)

(3) Electrical units are rationalized and are obtained by assigning the value $4\pi \times 10^{-7}\ \mathrm{H\ m^{-1}}$ to μ_0, the permeability of a vacuum. This leads to the ampere - one of the seven base units - as the unit of current. The other electrical units are derived directly from the base units and are identical with the practical units. (See Duffin 1980 for a clear discussion of electrical units.)

(4) Multiples and fractions of units are normally restricted to powers of 1000. So the centimetre is frowned on in strict SI. On the other hand, the step between $1\ \mathrm{mm^3}$ and $1\ \mathrm{m^3}$ is so large that the litre ($= 10^{-3}\ \mathrm{m^3}$) is retained as a convenient unit of volume.

SI units – names and symbols

Quantity	Unit	Symbol	Relation to other units
Base units			
length	metre	m	
mass	kilogram	kg	
time	second	s	
electric current	ampere	A	
thermodynamic temperature	kelvin	K	
amount of substance	mole	mol	
luminous intensity	candela	cd	
Supplementary units			
plane angle	radian	rad	
solid angle	steradian	sr	
Derived units with special names			
force	newton	N	$kg\,m\,s^{-2}$
pressure	pascal	Pa	$N\,m^{-2}$
energy	joule	J	N m
power	watt	W	$J\,s^{-1}$
electric charge	coulomb	C	A s
potential	volt	V	$J\,C^{-1}$
resistance	ohm	Ω	$V\,A^{-1}$
capacitance	farad	F	$C\,V^{-1} = s\,\Omega^{-1}$
magnetic flux	weber	Wb	V s
flux density	tesla	T	$Wb\,m^{-2}$
inductance	henry	H	$V\,s\,A^{-1} = \Omega\,s$
frequency	hertz	Hz	s^{-1}
temperature	degree Celsius	°C	$t/°C = T/K - 273{\cdot}15$
luminous flux	lumen	lm	cd sr
illuminance	lux	lx	$lm\,m^{-2}$
activity (radioactive)	becquerel	Bq	s^{-1}
absorbed dose (of ionizing radiation)	gray	Gy	$J\,kg^{-1}$

Decimal fractions and multiples

Fraction	Prefix	Symbol	Multiple	Prefix	Symbol
10^{-3}	milli	m	10^3	kilo	k
10^{-6}	micro	μ	10^6	mega	M
10^{-9}	nano	n	10^9	giga	G
10^{-12}	pico	p	10^{12}	tera	T
10^{-15}	femto	f	10^{15}	peta	P
10^{-18}	atto	a	10^{18}	exa	E

Note. The prefix is to be taken together with the unit *before* the power symbol operates, e.g.,

$$1\ \mu\text{m}^2 = 1\ (\mu\text{m})^2 = 10^{-12}\ \text{m}^2.$$

Relation to c.g.s. units

Electrical units

$1\ \text{A} = 10^{-1}$ e.m.u. $= c/10$ e.s.u.
$1\ \text{V} = 10^8$ e.m.u. $= 10^8/c$ e.s.u.
$1\ \Omega = 10^9$ e.m.u. $= 10^9/c^2$ e.s.u.
$1\ \text{F} = 10^{-9}$ e.m.u. $= 10^{-9}\ c^2$ e.s.u.
$1\ \text{H} = 10^9$ e.m.u. $= 10^9/c^2$ e.s.u.
$1\ \text{T} = 10^4$ e.m.u. (gauss)
$1\ \text{Wb} = 10^8$ e.m.u. (maxwell)
$1\ \text{A m}^{-1} = 4\pi \times 10^{-3}$ e.m.u. (oersted)
($c \approx 3 \times 10^{10}$ in the above equations.)

Other units

length	1 micron $= 10^{-6}$ m
	1 ångström $= 10^{-10}$ m
	1 X unit $= 10^{-13}$ m
	1 fermi $= 10^{-15}$ m
area	1 barn $= 10^{-28}\ \text{m}^2$
force	1 dyne $= 10^{-5}$ N
energy	1 erg $= 10^{-7}$ J
	1 calorie (IT) $= 4{\cdot}1868$ J
	1 electron volt $= 1{\cdot}6022 \times 10^{-19}$ J
pressure	1 bar $= 10^5$ Pa
	1 atmosphere $= 1{\cdot}01325 \times 10^5$ Pa
	1 torr $= 1$ mm of Hg $= 133{\cdot}322$ Pa
viscosity	
dynamic	1 poise $= 10^{-1}$ Pa s
kinematic	1 stokes $= 10^{-4}\ \text{m}^2\ \text{s}^{-1}$

Definitions of the SI base units

Metre

The *metre* is the length of the path travelled by light in vacuum during a time interval of 1/299 792 458 of a second.

Kilogram

The *kilogram* is the unit of mass; it is equal to the mass of the international prototype of the kilogram.

Second

The *second* is the duration of 9 192 631 770 periods of the radiation corresponding to the transition between the two hyperfine levels of the ground state of the caesium-133 atom.

Ampere

The *ampere* is that constant current which, if maintained in two straight parallel conductors of infinite length, of negligible circular cross-section, and placed 1 metre apart in vacuum, would produce between these conductors a force equal to 2×10^{-7} newton per metre of length.

Kelvin

The *kelvin*, unit of thermodynamic temperature, is the fraction 1/273·16 of the thermodynamic temperature of the triple point of water.

Mole

The *mole* is the amount of substance of a system which contains as many elementary entities as there are atoms in 0·012 kilogram of carbon-12. When the mole is used, the elementary entities must be specified and may be atoms, molecules, ions, electrons, other particles, or specified groups of such particles.

Candela

The *candela* is the luminous intensity, in the perpendicular direction, of a surface of 1/600 000 square metre of a black body at the temperature of freezing platinum under a pressure of 101 325 newtons per square metre.

The above definition of the metre came into effect in 1983 and replaced the definition which assigned a certain value to the wavelength of one of the lines in the spectrum of the krypton-86 atom. The new definition means that the speed of light is now *defined* to be exactly 299 792 458 m s^{-1}. For details and discussion of the new definition see Petley 1983 and *Metrologia* **19**, 163, 1984. For further details of the other units see Kaye and Laby 1973, p. 6.

APPENDIX G

Values of physical constants

speed of light	$c = 2{\cdot}9979 \times 10^{8}\ \mathrm{m\ s^{-1}}$
permittivity of vacuum	$\varepsilon_0 = 1/\mu_0 c^2 = 8{\cdot}8542 \times 10^{-12}\ \mathrm{F\ m^{-1}}$
elementary charge	$e = 1{\cdot}6022 \times 10^{-19}\ \mathrm{C}$
Boltzmann constant	$k = 1{\cdot}3807 \times 10^{-23}\ \mathrm{J\ K^{-1}}$
Avogadro constant	$N_A = 6{\cdot}0221 \times 10^{23}\ \mathrm{mol^{-1}}$
Faraday constant	$F = N_A e = 9{\cdot}6485 \times 10^{4}\ \mathrm{C\ mol^{-1}}$
gas constant	$R = N_A k = 8{\cdot}3145\ \mathrm{J\ K^{-1}\ mol^{-1}}$
atomic mass unit	$m_u = 10^{-3}/N_A = 1{\cdot}6605 \times 10^{-27}\ \mathrm{kg}$
energy equivalent	$m_u c^2 = 931{\cdot}49\ \mathrm{MeV}$
mass of electron	$m_e = 9{\cdot}1094 \times 10^{-31}\ \mathrm{kg}$
mass of proton	$m_p = 1{\cdot}6726 \times 10^{-27}\ \mathrm{kg}$
mass of neutron	$m_n = 1{\cdot}6749 \times 10^{-27}\ \mathrm{kg}$
Planck constant	$h = 6{\cdot}6261 \times 10^{-34}\ \mathrm{J\ s}$
	$\hbar = h/2\pi = 1{\cdot}0546 \times 10^{-34}\ \mathrm{J\ s}$
Rydberg constant	$R_\infty = 1{\cdot}0974 \times 10^{7}\ \mathrm{m^{-1}}$
fine structure constant α	$1/\alpha = 137{\cdot}036$
Bohr magneton	$\mu_B = 9{\cdot}2740 \times 10^{-24}\ \mathrm{J\ T^{-1}}$
nuclear magneton	$\mu_N = 5{\cdot}0508 \times 10^{-27}\ \mathrm{J\ T^{-1}}$
standard volume of ideal gas	$V_0 = 22{\cdot}414 \times 10^{-3}\ \mathrm{m^3\ mol^{-1}}$
Stefan–Boltzmann constant	$\sigma = 5{\cdot}671 \times 10^{-8}\ \mathrm{W\ m^{-2}\ K^{-4}}$
gravitational constant	$G = 6{\cdot}673 \times 10^{-11}\ \mathrm{N\ m^2\ kg^{-2}}$
acceleration due to gravity	

$$g = (9{\cdot}7803 + 0{\cdot}0519 \sin^2\phi - 3{\cdot}1 \times 10^{-6}\, H)\ \mathrm{m\ s^{-2}}$$

ϕ = latitude; H = height above sea level in metres

The values of the physical constants, other than g, are taken from Cohen and Taylor 1987; the quoted error is 1 or less in the last digit. The expression for g is taken from Kaye and Laby 1986, p. 159, and gives values correct to about $5 \times 10^{-4}\ \mathrm{m\ s^{-2}}$.

SOLUTIONS TO EXERCISES

3.1 The value of the mean is 9·803 m s^{-2}, and the values of the residuals in units of 10^{-2} m s^{-2} are 1, −1, 4, 1, −5, −1, 3, giving $\sum d_i^2 = 54$. Dividing this by $n-1=6$, and taking the square root, gives $\sigma = 0{\cdot}030$ m s^{-2}. $\sigma_m = 0{\cdot}030/\sqrt{7} = 0{\cdot}011$ m s^{-2}. The group result is thus $g = 9{\cdot}803 \pm 0{\cdot}011$ m s^{-2}.

If you have a calculator programmed to calculate standard deviations, try feeding in the values of g, and check that you get the same value for σ. Note that the calculator gives σ, and not σ_m. Some calculators are programmed to calculate s (equation 3.14), rather than σ, i.e. the quantity $\sum d_i^2$ is divided by n instead of $n-1$. You can check your own calculator by feeding in the numbers 9 and 11, for which $\sigma = \sqrt{2}$ and $s = 1$.

3.2 The values of E and the standard errors are given in units of 10^{11} N m^{-2}.
Newton's rings experiment. The mean value of E is 1·98. The value of σ given by the standard method is 0·25. The range method gives $\sigma \approx 0{\cdot}68/\sqrt{10} = 0{\cdot}22$.
Dial indicator experiment. The mean value of E is 2·047. The estimates of σ are 0·028 (standard method) and 0·025 (range method).

Dividing σ by $\sqrt{10}$ to obtain σ_m we have

Newton's rings experiment $E = 1{\cdot}98 \pm 0{\cdot}08$ (or 0·07)
Dial indicator experiment $E = 2{\cdot}047 \pm 0{\cdot}009$ (or 0·008).

The difference between the two mean values is slightly less than the standard error in the Newton's rings value, so there is little evidence for a systematic difference between the two experimental methods.

3.3 (a) 0·00266, (b) 0·0161, (c) 0·00036,
(d) 0·683, (e) 0·954, (f) 0·997.

The fraction of readings between x and $x + \mathrm{d}x$ is $f(z)\,\mathrm{d}z$, where $f(z)$, is defined on p. 31 and tabulated in Appendix A; $z = x/\sigma$. For exercises (a) to (c) $\mathrm{d}z = 0{\cdot}1/15{\cdot}0$. So the answer to (a) is $0{\cdot}399/150 = 0{\cdot}00266$. The answers to (d), (e), (f) are given by the values of $\phi(z)$ for $z = 1, 2, 3$.

4.1 In the following solutions, $g(A)$ means the % standard error in A, i.e.

$$g(A) = 100\,\Delta A/A.$$

(a)
$$g(A) = 4, \qquad g(A^2) = 8,$$
$$Z = 625, \qquad \Delta Z = 625 \times 8/100 = 50,$$
$$\underline{Z = 625 \pm 50.}$$

(b)
$$2B = 90 \pm 4, \qquad \Delta Z = (3^2 + 4^2)^{\frac{1}{2}} = 5,$$
$$\underline{Z = 10 \pm 5.}$$

(c)
$$g(C) = 1, \qquad g(C^2) = 2, \qquad C^2 = 2500 \pm 50,$$
$$g(D) = 8, \qquad g(D^{\frac{3}{2}}) = 12, \qquad D^{\frac{3}{2}} = 1000 \pm 120.$$

Put $$E = C^2 + D^{\frac{3}{2}} = 3500.$$

$$\Delta E = (50^2 + 120^2)^{\frac{1}{2}} = 130, \qquad g(E) = 3{\cdot}7,$$

$$g(A) = 3, \qquad g(B) = 5,$$

$$g(Z) = (3^2 + 5^2 + 3{\cdot}7^2)^{\frac{1}{2}} = 6{\cdot}9,$$

$$\underline{Z = 350 \pm 24.}$$

(d) $$\frac{\Delta(\ln B)}{\ln B} = \frac{\Delta B/B}{\ln B} = \frac{0{\cdot}02}{4{\cdot}61} = \frac{0{\cdot}43}{100},$$

$$g(Z) = (0{\cdot}6^2 + 0{\cdot}43^2)^{\frac{1}{2}} = 0{\cdot}74,$$

$$\underline{Z = 46{\cdot}1 \pm 0{\cdot}3.}$$

(e) $$g(A) = 4, \qquad g\left(\frac{1}{A}\right) = 4,$$

$$\frac{1}{A} = 0{\cdot}0200 \pm 0{\cdot}0008,$$

$$\underline{Z = 0{\cdot}9800 \pm 0{\cdot}0008.}$$

4.2 (a) The measured values of l_x, l_y, l_z are independent. From Table 4.1 (ii), the standard error in the volume is therefore

$$\frac{\sqrt{3}}{100} \approx 0{\cdot}02\%.$$

(b) The measured values are not independent. An increase in temperature causes all three lengths to increase by the same fractional amount. The situation is represented by (4.9) with $n = 3$. The standard error in the volume is therefore $0{\cdot}03\%$.

In practice, variations in the measured values of the length could be due to both instrumental errors and temperature fluctuations. It would be necessary to decide the contributions of each in order to estimate the error in the volume.

4.3 The values and standard errors of the slope are:

Method	Slope/μm kg^{-1}
least squares	$-349{\cdot}2 \pm 1{\cdot}9$
points in pairs	$-350{\cdot}1 \pm 2{\cdot}0$

The calculation by the method of least squares is made easier if the values of x are taken to be $4W$, where W is in kg. Then $\bar{x}$ and all the $(x_i - \bar{x})$ are simple integers.

4.4 The values and standard errors of the slope are

Method	Slope/mV K^{-1}
least squares	
errors only in V	$2{\cdot}551 \pm 0{\cdot}041$
errors only in T	$2{\cdot}556 \pm 0{\cdot}041$
points in pairs	$2{\cdot}542 \pm 0{\cdot}053$

The values are given to 3 decimal places for purposes of comparison, but normally these results would be quoted to only 2 decimal places.

Notice that the results of the two least-squares calculations are quite close. In fact there are probably errors in both the voltage and temperature readings, but a calculation that takes this into account requires a knowledge of the relative errors in V and T. The value of the slope of the best line when there are errors in both variables always lies between the two values corresponding to errors in only one or other of the variables (Guest 1961, p. 131). Since these two values are usually quite close, calculations are usually done on the assumption that only one of the variables is subject to error.

4.5 Express each value of the mass as

$$m = (139\,560 + x) \pm \Delta x \text{ keV}.$$

Weight each value inversely as the square of its standard error.

x	Δx	$w = (10/\Delta x)^2$	wx
9	8	2	18
11	10	1	11
8·6	2·0	25	215
6·7	2·4	17	114
5·8	1·8	31	180
7·5	0·9	123	923

$$\bar{x} = \frac{\sum wx}{\sum w} = \frac{1461}{199} = 7{\cdot}3.$$

To calculate the standard error in the weighted mean we note that a weight of 123 corresponds to an error of 0·9. So a weight of 199 corresponds to an error of

$$\left[\frac{123}{199}\right]^{\frac{1}{2}} \times 0{\cdot}9 = 0{\cdot}7.$$

Thus $m = 139\,567{\cdot}3 \pm 0{\cdot}7$ keV.

5.1 (a)

$$\rho = \frac{M}{abc}.$$

The fractional error in ρ due to each measured quantity is equal to the fractional error in the quantity. The fractional error in b - 10% - entirely dominates the others. The error in ρ is therefore 10%. (The term 'error' in this and following solutions refers to standard error.)

(b)

$$a^2 = 6400 \pm 160 \text{ mm}^2,$$
$$b^2 = 100 \pm 20 \text{ mm}^2.$$

The error in a^2 now dominates that of b^2. The error in $a^2 + b^2$ is 2·5%. The error in M is negligible by comparison. The error in I is therefore 2·5% and comes entirely from the error in a.

5.2 The error in ϕ/C is 3%. The error in r is 2%; so the error in r^4 is 8%. The error in l is negligible. The error in n is $(3^2 + 8^2)^{\frac{1}{2}} = 8{\cdot}5\%$. Therefore

$$n = (8{\cdot}0 \pm 0{\cdot}7) \times 10^{10} \text{ N m}^{-2}.$$

Notice that a quantity raised to a high power needs to be measured relatively precisely.

5.3

$$\mu x = \ln I_0 - \ln I = 0{\cdot}7829.$$

The simplest way of finding the error is that of direct substitution. Thus

$$\ln(I_0 + \Delta I_0) - \ln I_0 = 0{\cdot}006,$$
$$\ln(I + \Delta I) \quad - \ln I \quad = 0{\cdot}011.$$

These give a combined error of 0·012, i.e.

$$\mu x = 0{\cdot}783 \pm 0{\cdot}012.$$

$x = 10$ mm with negligible error. Therfore

$$\mu = 78{\cdot}3 \pm 1{\cdot}2\ \mathrm{m}^{-1}.$$

The expression for the error given by the formal method (section 4.1 (b)) is

$$(\Delta\mu)^2 = \frac{1}{x^2}\left[\left(\frac{\Delta I}{I}\right)^2 + \left(\frac{\Delta I_0}{I_0}\right)^2\right] + \left(\frac{\mu\,\Delta x}{x}\right)^2,$$

which is not difficult to evaluate. (Again the term in Δx is negligible.) But you have to be able to derive the expression correctly in the first place. You are less liable to make mistakes with the simpler method of direct substitution.

5.4 As in exercise 5.3 the error can be evaluated either by direct substitution or formally. Since n and d are fixed, $\lambda \propto \sin\theta$. Evaluate $\sin\theta$ for θ and $\theta + \Delta\theta$. The results are 0·1959 and 0·1985, giving $\Delta\lambda/\lambda = 1{\cdot}3\%$. Since $E \propto (\text{momentum})^2 \propto 1/\lambda^2$, $\Delta E/E = 2\Delta\lambda/\lambda = \underline{2{\cdot}6\%}$. The expression given by the formal method is $\Delta E/E = 2\cot\theta\,\Delta\theta$, where

$$\Delta\theta = \frac{9}{60} \times \frac{\pi}{180}\ \text{rad}.$$

5.5 We have

$$L \propto \frac{f^2}{E}.$$

Denote the fractional increases in f and E by r_f and r_E. Then the fractional increase in L is

$$r_L = 2r_f - r_E.$$

Since the rise in temperature is 10 K,

$$\begin{aligned}\alpha &= r_L/10\\ &= (-0{\cdot}500 + 0{\cdot}520) \times 10^{-2} \times 10^{-1}\\ &= \underline{20 \times 10^{-6}}.\end{aligned}$$

The error in $2r_f$ is 4×10^{-5}; the error in r_E is 3×10^{-5}. The error in r_L is therefore 5×10^{-5} and the error in α is $\underline{5 \times 10^{-6}}$.

The method is a bad one. The two primary quantities r_f and r_E are measured to better than 1 part in 100, but the value of α is only good to 1 part in 4. This is an example of Case II, p. 53.

6.1 (a) Let the interval between flashes be

$$T_0 = \frac{1}{f_0}.$$

Suppose that f is slightly greater than mf_0. Then in time T_0 the object will make m revolutions plus a small fraction δ of a revolution. The body is actually rotating with frequency

$$f = \frac{m+\delta}{T_0} = (m+\delta)f_0,$$

but it appears to be rotating with frequency

$$f_{\text{app}} = \frac{\delta}{T_0} = \delta f_0 = f - mf_0.$$

The sense of the apparent rotation is the same as that of the actual rotation. If f is slightly less than mf_0, the quantity δ is negative and therefore f_{app} is negative. The body appears to be rotating in the opposite sense to that of the actual rotation.

If f is exactly equal to mf_0 the body appears stationary. This can sometimes happen with fluorescent lighting, where the variation of illumination as the mains voltage varies sinusoidally is much more marked than with a filament bulb. For this reason it is dangerous to have fluorescent lighting in workshops with machinery rotating at high speeds.

(b)

$$f = mf_0 + f_{\text{app}},$$
$$mf_0 = 500{\cdot}00 \pm 0{\cdot}05 \text{ Hz},$$
$$f_{\text{app}} = 0{\cdot}40 \pm 0{\cdot}05 \text{ Hz}.$$

Therefore

$$f = 500{\cdot}40 \pm 0{\cdot}07 \text{ Hz}.$$

6.2 In the first method, ΔE is obtained by measuring E_1^{20} and E_1^{30}, the emf of cell S_1 at 20 °C and 30 °C.

$$\Delta E = E_1^{20} - E_1^{30}.$$

The value of ΔE is 370 μV. Since the error in each value of E_1 is 10 μV, the error in ΔE is $10\sqrt{2} = 14$ μV. So the error is 4%.

In the second method, the small difference e between the emfs of the two cells S_1 and S_0 is measured directly. Denote the emf of S_0 by E_0. Put

$$e^{20} = E_1^{20} - E_0,$$
$$e^{30} = E_1^{30} - E_0.$$

Then

$$\Delta E = e^{20} - e^{30}.$$

The fractional error in each e is the same as that of the current through the milliammeter, subject to the condition that the absolute error cannot be less than the smallest voltage the galvanometer can detect. Suppose that $e^{20} = -10$ μV. (Provided it is small compared with ΔE, its actual magnitude does not affect the conclusion.) Then

$$e^{20} = -10 \pm 0{\cdot}2 \ \mu\text{V},$$
$$e^{30} = -380 \pm 4 \ \mu\text{V},$$
$$\Delta E = 370 \pm 4 \ \mu\text{V},$$

and the error is 1%.

The second method is therefore superior. The fractional change in E_1 with temperature is small. By subtracting a constant quantity E_0 - almost equal to E_1 - we obtain a quantity e whose fractional change with temperature is large. Notice that we do not need to know the value of E_0 - it is only necessary that it remain constant.

6.3 For a comprehensive account of the measurement of temperature see Quinn 1983.

6.4 See Germain 1963, Swithenby 1974, and Putley 1975.

6.8 An extremely readable account of the concept and measurement of extremes of temperature has been given by Zemansky 1964.

7.1 Denote the emfs of E and G by V_E and V_G and the current through them by I_E and I_G. Then

$$I_G S = V_G - V_E,$$
$$(I_G + I_E) R = V_E.$$

Therefore

$$\frac{S}{R} = \frac{(V_G/V_E) - 1}{1 - [I_E/(I_G + I_E)]} = \frac{v - 1}{1 - u}.$$

7.2 For a sinusoidal voltage of angular frequency $\omega = 2\pi f$, the impedance of the capacitor is $1/\mathrm{j}\omega C$. $\mathrm{j} = \sqrt{(-1)}$. Therefore

$$\frac{V_C}{V_Q} = \frac{1/\mathrm{j}\omega C}{(1/\mathrm{j}\omega C) + R} = \frac{1}{1 + \mathrm{j}\omega CR}.$$

$$\left|\frac{V_C}{V_Q}\right|^2 = \frac{1}{(1 + \mathrm{j}\omega CR)(1 - \mathrm{j}\omega CR)} = \frac{1}{1 + \omega^2 C^2 R^2},$$

which gives the required result.

7.3

$$\Delta V_Z = (\mathrm{d}V_Z/\mathrm{d}I_Z)\Delta I_Z$$
$$= 3 \times 0{\cdot}02 \times 10^{-3} = 6 \times 10^{-5}\ \mathrm{V}.$$

Therefore $\Delta V_Z / V_Z = 10^{-5}$.

7.4 Outside the Earth's surface, g is proportional to $1/R^2$, where R is the distance from the centre of the Earth. Therefore a decrease in g of Δg corresponds to an increase in height Δh, given by

$$\frac{\Delta g}{g} = 2\frac{\Delta h}{R_E}.$$

where R_E, the radius of the Earth, is 6400 km. Thus

$$\Delta h = \tfrac{1}{2} \times 6{\cdot}4 \times 10^6 \times 6 \times 10^{-9}\ \mathrm{m} = 19\ \mathrm{mm}.$$

7.5 A good description of this instrument and the method of exact fractions is to be found in Longhurst 1973, p. 185.

7.6 See Sproull and Phillips 1980, p. 31. The doublet method, like the stroboscope, is an example of Case I, p. 53.

12.1 The values used for the atomic constants in these solutions are approximate ones that you will find it useful to remember. In fact the values of these constants are probably known to a few parts in 10^6.

(a) 60 W.
The thermal conductivity of copper at 0 °C is 385 W m^{-1} K^{-1}. The calculation assumes no heat loss along the rod.

(b) The readings are too low by about 0·010%. The linear expansivity is about 10 to 11×10^{-6} for most kinds of steel.

(c) (i) 0·0199 Ω. (ii) It increases by 0·0017 Ω.
The resistivity of copper is $1{\cdot}56 \times 10^{-8}$ Ω m at 0 °C and it increases by 0·4% per degree rise in temperature.

(d) 4·1 mV.

(e) (i) 0·25 m s^{-1}. (ii) 0·45 m s^{-1}.
At 20 °C the viscosity of water is $1{\cdot}00 \times 10^{-3}$ N s m^{-2}, and at 50 °C it is $0{\cdot}55 \times 10^{-3}$ N s m^{-2}.

(f) 13 kN.
The Young modulus of steel is $2{\cdot}1 \times 10^{11}$ N m^{-2}.

(g) 1·29 m.
The speed of sound in air at 0 °C is 331 m s^{-1}.

(h) 1·9 km s^{-1}.
The result may be obtained from the equation

$$\tfrac{1}{2}mv^2 = \tfrac{3}{2}kT.$$

m, the mass of the hydrogen molecule is $2 \times 1{\cdot}67 \times 10^{-27}$ kg, and k, the Boltzmann constant, is $1{\cdot}38 \times 10^{-23}$ J K^{-1}.

(i) $G = 6{\cdot}7 \times 10^{-11}$ N m^2 kg^{-2}.
Use the equation

$$g = \frac{GM}{R_E^2},$$

where M is the mass and R_E the radius of the Earth. The mean density of the Earth is 5500 kg m^{-3} and its radius is 6400 km. $g = 9{\cdot}81$ m s^{-2}.

(j) Typical wavelengths for red, green, and violet light are 700 nm, 550 nm, and 400 nm. These values give (i) 710 lines per mm, (ii) 23·1°, (iii) 16·6°.

(k) 18 W.
Stefan–Boltzmann constant $\sigma = 5{\cdot}67 \times 10^{-8}$ W m^{-2} K^{-4}.

(l) (i) $1{\cdot}9 \times 10^7$ m s^{-1}. (ii) 39 pm.

mass of electron	$m_e = 9{\cdot}1 \times 10^{-31}$ kg,
elementary charge	$e = 1{\cdot}6 \times 10^{-19}$ C,
Planck constant	$h = 6{\cdot}6 \times 10^{-34}$ J s.

(m) 0·29 T.

mass of proton $\quad m_p = 1{\cdot}67 \times 10^{-27}$ kg.

(n) 91 nm.

The answer is the reciprocal of the Rydberg constant

$$R = 1{\cdot}10 \times 10^7 \text{ m}^{-1}.$$

(o) 931 MeV.

The mass in kilograms of a particle of atomic weight unity is $10^{-3}/N_A$, where $N_A = 6{\cdot}0 \times 10^{23}$ is the Avogadro constant.

$$\text{speed of light} \qquad c = 3{\cdot}0 \times 10^8 \text{ m s}^{-1}.$$

12.2 (a)

$$(1+\delta_1)(1+\delta_2)(1+\delta_3) \approx 1+\delta_1+\delta_2+\delta_3,$$

if the δs are small compared with 1. So the answer is

$$1 + 0{\cdot}000\,25 + 0{\cdot}000\,41 - 0{\cdot}000\,13 = \underline{1{\cdot}000\,53.}$$

(b)

$$\frac{1}{(1+\delta)^2} \approx 1 - 2\delta \quad \text{if} \quad \delta \ll 1.$$

δ is approximately 9/72 000, so 912·64 must be reduced by 18 parts in 72 000, that is, by 1 part in 4000, which is 0·23. The answer is <u>912·41.</u>

(c)

$$\begin{aligned}(9{\cdot}100)^{\frac{1}{2}} &= 3 \times (1 + \tfrac{1}{90})^{\frac{1}{2}} \\ &\approx 3 \times (1 + \tfrac{1}{180}) \\ &= \underline{3{\cdot}017.}\end{aligned}$$

SOME USEFUL BOOKS

Techniques and experimental methods

Carpenter, L. G. 1983. *Vacuum Technology*, 2nd ed., Adam Hilger.

Jones, B. E. 1982. Editor, *Instrument Science and Technology*, Adam Hilger.

Marton, L. 1959. Editor-in-chief, *Methods of Experimental Physics*, Academic Press. (Since 1983 the Editors-in-chief of this series have been Celotta, R. and Levine, J.)

O'Handon, J. F. 1980. *A User's Guide to Vacuum Technology*, John Wiley and Sons.

Electronics and instrumentation

Clayton, G. B. 1979. *Operational Amplifiers*, 2nd ed., Butterworths.

de Sa, A. 1981. *Principles of Electronic Instrumentation*, Arnold.

Horowitz, P. and Hill, W. 1980. *The Art of Electronics*, Cambridge University Press.

Malmstadt, H. V., Enke, C. G., and Crouch, S. R. 1981. *Electronics and Instrumentation for Scientists*, Benjamin/Cummings.

Millman, J. 1979. *Microelectronics*, McGraw-Hill.

Feedback and control theory

Schwarzenbach, J. and Gill, K. F. 1984. *System Modelling and Control*, 2nd ed., Arnold.

Shinners, S. M. 1978. *Modern Control System Theory and Application*, 2nd ed., Addison-Wesley.

Microprocessors and microcomputers

Garrett, P. H. 1978. *Analog Systems for Microprocessors and Minicomputers*, Prentice-Hall.

Klingman, E. E. 1977 and 1982. *Microprocessor Systems Design*, Vols. I and II, Prentice-Hall.

Morris, N. M. 1983. *Logic Circuits*, 3rd ed., McGraw-Hill.

Zaks, R. and Lesea, A. 1979. *Microprocessor Interfacing Techniques*, 3rd revised Ed., Sybex.

Mathematical tables

Abramowitz, M. and Stegun, I. A. 1964. *Handbook of Mathematical Functions*, National Bureau of Standards, reprinted by Dover Publications, 1970.

Scientific writing

Dampier, W. C. and Dampier, M. 1924. Editors, *Readings in the Literature of Science*, Cambridge University Press, reprinted by Harper & Brothers, 1959.

Fowler, H. W. 1965. *Modern English Usage*, 2nd ed., Oxford University Press.

Quiller-Couch, A. 1916. *The Art of Writing*, Cambridge University Press.

REFERENCES

Baer, T., Kowalski, F. V., and Hall, J. L. 1980. *Applied Optics*, **19**, 3173.

Baird, K. M. 1983. *Physics Today*, **36**, 52.

Bay, Z., Luther, G. G., and White, J. A. 1972. *Phys. Rev. Letters*, **29**, 189.

Bay, Z. and White, J. A. 1972. *Phys. Rev.*, **D5**, 796.

Birge, R. T. and Menzel, D. H. 1931. *Phys. Rev.*, **37**, 1669.

Bomford, G. 1980. *Geodesy*, 4th ed., Clarendon Press, Oxford.

Braginsky, V. B. and Panov, V. I. 1972. *Sov. Phys. JETP*, **34**, 463.

Candler, C. 1951. *Modern Interferometers*, Hilger & Watts.

Cohen, E. R. and DuMond, J. W. M. 1965. *Rev. Mod. Phys.*, **37**, 537.

Cohen, E. R. and Taylor, B. N. 1987. *Rev. Mod. Phys.*, **59**, 1121.

Cook, A. H. 1967. *Contemporary Physics*, **8**, 251.

Cook, A. H. 1973. *Physics of the Earth and Planets*, Macmillan.

Cook, A. H. 1975, *Contemporary Physics*, **16**, 395.

Deslattes, R. D., Henins, A., Bowman, H. A., Schoonover, R. M., Carroll, C. L., Barnes, I. L., Machlan, L. A., Moore, L. J., and Shields, W. R. 1974. *Phys. Rev. Letters*, **33**, 463.

Dicke, R. H. 1961. *Scientific American*, **205**, 6, 84.

Duffin, W. J. 1980. *Electricity and Magnetism*, 3rd ed., McGraw-Hill.

Evenson, K. M., Wells, J. S., Petersen, F. R., Danielson, B. L., and Day, G. W. 1973. *Appl. Phys. Letters*, **22**, 192.

Fowler, R. H. 1936. *Statistical Mechanics*, 2nd ed., Cambridge University Press.

Friend, R. H. and Bett, N. 1980. *J. Phys. E.* **13**, 294.

Froome, K. D. 1958, *Proc. Roy. Soc.*, **A247**, 109.

Gallop, J. C. 1982. *Metrologia*, **18**, 67.

Germain, C. 1963. *Nuc. Inst. and Methods*, **21**, 17.

Gough, W., Richards, J. P. G., and Williams, R. P. 1983. *Vibrations and Waves*, Ellis Horwood.

Guest, P. G. 1961. *Numerical Methods of Curve Fitting*, Cambridge University Press.

Hecht, E. and Zajac, A. 1974. *Optics*, Addison-Wesley.

Horowitz, P. and Hill, W. 1980. *The Art of Electronics*, Cambridge University Press.

Kappler, E. 1938. *Ann. d. Physik*, 5, **31**, 377.

Kartaschoff, P. 1978. *Frequency and Time*, Academic Press.

Kaye, G. W. C. and Laby, T. H. 1986. *Tables of Physical and Chemical Constants*, 15th ed., Longman.

Longhurst, R. S. 1973. *Geometrical and Physical Optics*, 3rd ed., Longman.

Luther, G. G. and Towler, W. R., 1982. *Phys. Rev. Letters*, **48**, 121.

Millman, J. 1979. *Microelectronics*, McGraw-Hill.

Petley, B. W. 1971. *Contemporary Physics*, **12**, 453.

Petley, B. W. 1980. *Contemporary Physics*, **21**, 607.
Petley, B. W. 1983. *Nature*, **303**, 373.
Petley, B. W. 1985. *The Fundamental Physical Constants and the Frontier of Measurement*, Adam Hilger.
Pound, R. V. and Rebka, G. A. 1960. *Phys. Rev. Letters*, **4**, 337.
Putley, E. H. 1975. *Contemporary Physics*, **16**, 101.
Quinn, T. J. 1983. *Temperature*, Academic Press.
Rayleigh, Lord and Ramsay, W. 1895. *Phil. Trans. Roy. Soc.*, **186**, 187.
Reif, F. 1965. *Fundamentals of Statistical and Thermal Physics*, McGraw-Hill.
Reynolds, O. 1883. *Phil. Trans. Roy. Soc.*, **174**, 935.
Robinson, F. N. H. 1974. *Noise and Fluctuations in Electronic Devices and Circuits*, Clarendon Press, Oxford.
Rowan-Robinson, M. 1981. *Cosmology*, 2nd ed., Clarendon Press, Oxford.
Smith, C. J. 1960. *General Properties of Matter*, 2nd ed., Arnold.
Sproull, R. L. and Phillips, W. A. 1980. *Modern Physics*, 3rd ed., John Wiley & Sons.
Swithenby, S. J. 1974. *Contemporary Physics*, **15**, 249.
Sydenham, P. H. 1985. *Transducers in Measurement and Control*, 3rd ed., Adam Hilger.
Thomson, J. J. 1897. *Phil. Mag.*, *Ser.* 5, **44**, 293.
Urey, H. C., Brickwedde, F. G., and Murphy, G. M. 1932. *Phys. Rev.*, **39**, 864.
van der Pauw, L. J. 1958. *Philips Res. Rep.*, **13**, 1.
van der Ziel, A. 1976. *Noise in Measurements*, John Wiley & Sons.
Whittaker, E. and Robinson, G. 1944. *The Calculus of Observations*, 4th ed., Blackie & Son.
Williams, E. R., Faller, J. E., and Hill, H. A. 1971. *Phys. Rev. Letters*, **26**, 721.
Wohl, C. G. *et al.* 1984. *Rev. Mod. Phys.* **56**, *S*1.
Wood, A. 1940. *Acoustics*, Blackie & Son.
Zemansky, M. W. 1964. *Temperatures Very Low and Very High*, Van Nostrand.
Zumberge, M. A., Rinker, R. L., and Faller, J. E. 1982. *Metrologia*, **18**, 145.

INDEX